Environmental Chemistry for a Sustainable World

Volume 33

Other Publications by the Editors

Books

Environmental Chemistry
http://www.springer.com/978-3-540-22860-8

Organic Contaminants in Riverine and Groundwater Systems
http://www.springer.com/978-3-540-31169-0

Sustainable Agriculture
Volume 1: http://www.springer.com/978-90-481-2665-1
Volume 2: http://www.springer.com/978-94-007-0393-3

Book series

Environmental Chemistry for a Sustainable World
http://www.springer.com/series/11480

Sustainable Agriculture Reviews
http://www.springer.com/series/8380

Journals

Environmental Chemistry Letters
http://www.springer.com/10311

More information about this series at http://www.springer.com/series/11480

Anish Khan • Inamuddin • Abdullah M. Asiri
Editors

E-waste Recycling and Management

Present Scenarios and Environmental Issues

Springer

Editors
Anish Khan
Chemistry Department
King Abdulaziz University
Jeddah, Saudi Arabia

Inamuddin
Chemistry Department,
Faculty of Science
King Abdulaziz University
Jeddah, Saudi Arabia

Abdullah M. Asiri
Chemistry Department,
Faculty of Science
King Abdulaziz University
Jeddah, Saudi Arabia

ISSN 2213-7114 ISSN 2213-7122 (electronic)
Environmental Chemistry for a Sustainable World
ISBN 978-3-030-14183-7 ISBN 978-3-030-14184-4 (eBook)
https://doi.org/10.1007/978-3-030-14184-4

This Springer imprint is published by the registered company Springer Nature Switzerland AG.
The registered company address is: Gewerbestrasse 11, 6330 Cham, Switzerland

Preface

Expanding the rate of market infiltration, financial development and fast mechanical progression prompt the monstrous age of electronic waste in the habitat. The absence of awareness and coarse treatment of e-waste causes a worldwide threat. E-waste is a famous casual name for electronic items generated after their helpful life. Computers, televisions, video cameras, recorders, stereos, copiers and fax machines are some of the regular electronic items being generated as e-waste. A number of items may be reused after giving a face-lift. However, with quick changing advancements and consistent buyer interest for the most recent gadgets, the ascent in e-waste looks set to proceed. Indeed, the purchasers may play an essential role by picking items that are less perilous and are intended for more secure reusing. On the other hand, the e-waste management networks may be set up to guarantee the collection of inappropriately surrendered or censured items and outdated or broken electrical or electronic gadgets. It is obvious that the e-waste items, which are generated and not managed properly, contaminate nature, making it inadmissible for human home. On the other hand, the progression of advancements has decreased the life cycle of electronic items. Thus, the rate to outdated old items increases progressively. Thus, it is very necessary to be aware of the e-waste world and its impact on the global habitat of the species. In view of that, this book is planned to provide a comprehensive literature on the global e-waste recycling and management. This book is divided into 12 chapters as given below:

Chapter 1 explores the status of a cathode-ray tube disposal and environmental issues followed by potential challenges of segregating funnel and panel glass of cathode-ray tube. Separation of funnel and panel glass from the cathode-ray tube based on open-loop and closed-loop process is discussed with pros and cons.

Chapter 2 includes a description of methods of disassembly focused on e-waste recycling in compliance with environmental standards. The required steps of the end-of-life products disassembly vary depending on the category of waste equipment. To show these differences, the chapter includes two case studies showing the configuration of a layout of e-waste processing lines with possible options to reconfigure them. The variants of the system's configuration depend on the volume of the waste stream, labour cost and required purity of output materials. Economic efficiency indicator of e-waste is presented in this chapter on cooling appliances recycling for four European countries.

Chapter 3 discusses some of the most important factors, including legal, statistical, economic and organizational factors that affect the recycling of waste electrical and electronic equipment or more broadly the recycling of general electronic waste in Japan and other countries. The policy importance of incorporating manufacturing supply chains in the design of environmental management of production systems is emphasized. This chapter puts forward some recommendations that need to be taken into account in the public policy debate in order to improve the current low rates.

Chapter 4 discusses the current state of electronic waste management through technology. It begins by giving the definition and classification of electronic waste separation and recycling strategies. It is also mentioning the importance of electronic waste management and statistics of the exponential increase of electronic waste. After that, electronic waste is classified, and the major challenges faced in electric and electronic waste management and control regulations are discussed. Finally, the material composition in waste electrical and electronic equipment and current as well as future electronic waste management technologies are discussed in details.

Chapter 5 discusses the recycling challenges for the adoption of e-waste reverse logistics under the perspective of developing countries. It is also pointing out the categorization of the barriers in financial/economics; environmental; market related; legal; policy related; management; knowledge related and technical and technological related. The compilation of information related to recycling challenges of e-waste in developing countries and the identification of some solutions and actions to overcome these barriers are also discussed which can be useful for practitioners and researchers.

Chapter 6 explores the systematic methods used for the management of electronic waste. It provides information about electronic waste, plastics in electronic waste, electronic waste management issues, worldwide electronic waste generation and issues related to electronic waste and environmental public health. Finally, energy recovery from electronic waste using methods such as chemical recycling, mechanochemical treatment, hydrothermal process, pyrolysis, combustion process, gasification process, integrated process and hydrocracking is discussed.

Chapter 7 contributes to the literature on the management of waste electrical and electronic equipment (WEEE) by comparing the performance of the different European Union countries according to the targets set in the regulation of the Union's environmental policy on WEEE. To this end, the traditional non-parametric data envelopment analysis is used to measure technical efficiency for the first time in the literature. A sample of 30 European countries for the year 2014 is used with the purpose of comparing their performance, ranking the countries and identifying their level of inefficiency.

Chapter 8 addresses the various categories deployed towards effective e-waste management such as collection, disposal of dangerous portions and recovery of precious metals and energy. The benefits, challenges and future of e-waste management are also highlighted.

Chapter 9 discusses the methods used for the recycling of the precious metals obtained from the light-emitting diode industry. These metals are gallium, indium, rare earth elements like yttrium and cerium and precious metals such as gold and silver. Some of the most important methods developed for this purpose include pyrometallurgical (pyrolysis), hydrometallurgical (acid leaching) and biotechnological technologies (microbial leaching).

Chapter 10 discusses the current scenario in the electrical and electronic equipment industry and generation of waste electric and electronic equipment considering the implications of resource management and environment, social and economic impact in this production chain.

Chapter 11 deals with sustainable electronic waste management implications for environmental and human health. It is written to explain the electronic waste and sustainable development goals with electronic waste tracking and driving trends. The electronic waste statistics and measurement along the side positive and negative effects of electronic waste are also discussed. Some of the products that make challenges to a recycler are also discussed. Finally, the implications of electronic waste on human health and the environment discourse with the aim of electronic waste management are discussed.

Chapter 12 provides a brief insight into the global trends of e-waste generation, critical issues and challenges associated with e-waste and its effects on environmental and human health. Finally, the chapter highlights the need for sustainable environmental management of e-waste.

Jeddah, Saudi Arabia Inamuddin
Jeddah, Saudi Arabia Anish Khan
Jeddah, Saudi Arabia Abdullah M. Asiri

Contents

Contributors

Mohamed Aboughaly Faculty of Energy Systems and Nuclear Science, and Faculty of Engineering and Applied Science, University of Ontario Institute of Technology, Oshawa, ON, Canada

Jayaseelan Arun Department of Chemical Engineering, SSN College of Engineering, Kalavakkam, Tamil Nadu, India

D. Balaji Department of Chemical Engineering, SSN College of Engineering, Chennai, Tamil Nadu, India

Feliciane Andrade Brehm Universidade do Vale do Rio dos Sinos – UNISINOS, Porto Alegre, RS, Brazil

Lúcio Camara e Silva Management Engineering Department, CDSID – Center for Decision Systems and Information Development, Federal University of Pernambuco, Recife, Brazil

Gisele Lorena Diniz Chaves Department of Engineering and Technology, Federal University of Espírito Santo, Vitória, Espírito Santo, Brazil

Angéli Viviani Colling Universidade do Vale do Rio dos Sinos – UNISINOS, Porto Alegre, RS, Brazil

Tamires Augustin da Silveira Universidade do Vale do Rio dos Sinos – UNISINOS, Porto Alegre, RS, Brazil

Emanuele Caroline Araújo dos Santos Universidade do Vale do Rio dos Sinos – UNISINOS, Porto Alegre, RS, Brazil

Hossam A. Gabbar Faculty of Energy Systems and Nuclear Science, and Faculty of Engineering and Applied Science, University of Ontario Institute of Technology, Oshawa, ON, Canada

Kannappan Panchamoorthy Gopinath Department of Chemical Engineering, SSN College of Engineering, Kalavakkam, Tamil Nadu, India

K. Grace Pavithra Department of Chemical Engineering, SSN College of Engineering, Chennai, Tamil Nadu, India

Patricia Guarnieri Faculty of Economics, Administration, Accounting and Public Policies Management, University of Brasilia, Brasilia, Brazil

Hitoshi Hayami Faculty of Business and Commerce, Keio University, Tokyo, Japan

Victor C. Igbokwe Department of Microbiology, Faculty of Biological Sciences, University of Nigeria, Nsukka, Enugu State, Nigeria

Bioconversion and Renewable Energy Research Unit, University of Nigeria, Nsukka, Enugu State, Nigeria

Carlos Alberto Mendes Moraes Universidade do Vale do Rio dos Sinos – UNISINOS, Porto Alegre, RS, Brazil

Masao Nakamura Sauder School of Business, University of British Columbia, Vancouver, BC, Canada

Graduate School of Business and Commerce, Keio University, Tokyo, Japan

Isabel Narbón-Perpiñá Department of Business, Universitat Autònoma de Barcelona, Barcelona, Spain

Piotr Nowakowski Faculty of Transport, Silesian University of Technology, Katowice, Poland

Tochukwu N. Nwagu Department of Microbiology, Faculty of Biological Sciences, University of Nigeria, Nsukka, Enugu State, Nigeria

Charles O. Nwuche Department of Microbiology, Faculty of Biological Sciences, University of Nigeria, Nsukka, Enugu State, Nigeria

Joyce N. Odimba Department of Microbiology, Faculty of Biological Sciences, University of Nigeria, Nsukka, Enugu State, Nigeria

Bioconversion and Renewable Energy Research Unit, University of Nigeria, Nsukka, Enugu State, Nigeria

Chukwudi O. Onwosi Department of Microbiology, Faculty of Biological Sciences, University of Nigeria, Nsukka, Enugu State, Nigeria

Bioconversion and Renewable Energy Research Unit, University of Nigeria, Nsukka, Enugu State, Nigeria

Diego Prior Department of Business, Universitat Autònoma de Barcelona, Barcelona, Spain

Shahriar Shams Civil Engineering Programme Area, Universiti Teknologi Brunei, Bandar Seri Begawan, Brunei-Muara, Brunei Darussalam

Bhavisha Sharma Department of Environment and Sustainable Development, Institute of Environment and Sustainable Development, Banaras Hindu University, Varanasi, India

Pooja Singh Department of Science, Institute of Computer Science and Technology, SHEPA, Varanasi, India

Rajeev Pratap Singh Department of Environment and Sustainable Development, Institute of Environment and Sustainable Development, Banaras Hindu University, Varanasi, India

Panneer Selvam Sundar Rajan Department of Chemical Engineering, SSN College of Engineering, Chennai, Tamil Nadu, India

Barkha Vaish Department of Environment and Sustainable Development, Institute of Environment and Sustainable Development, Banaras Hindu University, Varanasi, India

Lúcia Helena Xavier Centre for Mineral Technology (CETEM), Rio de Janeiro, Brazil

Chapter 1
Solution and Challenges in Recycling Waste Cathode-Ray Tube

Shahriar Shams

Contents

Abstract The introduction of liquid-crystal display (LCD) for television and personal computer monitor has gained momentum in sales and distribution due to its portability and energy efficiency over traditional bulky cathode-ray tube (CRT) used in the manufacture of television and personal computer. The disposal of the

S. Shams (✉)
Civil Engineering Programme Area, Universiti Teknologi Brunei,
Bandar Seri Begawan, Brunei-Muara, Brunei Darussalam
e-mail: shams.shahriar@utb.edu.bn

A. Khan et al. (eds.), *E-waste Recycling and Management*,
Environmental Chemistry for a Sustainable World 33,
https://doi.org/10.1007/978-3-030-14184-4_1

cathode-ray tube is further having a major challenge due to its hazardous characteristics resulting from the composition of glass used in cathode-ray tubes. There are various recycling technologies available to extract toxic lead from funnel glass of cathode-ray tube. This chapter explores the status of cathode-ray tube, disposal, and environmental issues followed by potential challenges of segregating funnel and panel glass of cathode-ray tube. Separation of funnel and panel glass from the cathode-ray tube based on open-loop and closed-loop process is discussed with pros and cons. Cathode-ray tube glass-ceramic brick and concrete, vitrification glass to stabilize nuclear waste, and fluxing materials such as silica flux are gaining momentum on the reuse of cathode-ray tube under closed-loop process. The diamond cutting method for segregation of funnel glass from panel glass is highly recommended among the various potential segregation technologies available due to its vacuum adsorption and dust recovery capacity, automatic edge searching, and laser positioning. The study finds that emerging technology using furnace and chemicals for extraction of toxic lead from the cathode-ray tube is a promising method for management of recycling in an environmentally sustainable way without any residual waste.

Keywords Cathode-ray tube · Closed-loop process · Frit · Funnel glass · Lead · Open-loop process · Panel glass · Recycling · Television · Waste

1.1 Introduction

Cathode-ray tube (CRT) used as a display unit for television and computer monitor has become obsolete with the arrival of liquid-crystal display (LCD) screens, i.e., lead (Pb) in the cathode-ray tube is replaced by the introduction of mercury (Hg) in liquid-crystal display. Each cathode-ray tube contains several leaded glasses, which is quite heavy and requires special handling procedures for safety and regulations. Even when the glass is processed properly, its low value and other market limitations make it difficult to recycle into new products. The funnel and panel glass (Fig. 1.1) extracted from the old cathode-ray tube was extensively used to produce new cathode-ray tube for television and computer monitor using closed-loop recycling system almost 15 years ago. Hence, closed-loop recycling system segregated the glass from ferrous and nonferrous metals, oxides, phosphor, and dust efficiently and thus provided necessary assistance to the manufacturers of the cathode-ray tube in the past. The emergence of lightweight, space-saving, and energy-efficient light-emitting diode (LED) and plasma screens has significantly reduced the demand for new cathode-ray tube used in television and computer monitor, as cathode-ray tube consumes a lot more space due to its enormous size and heavyweight (He and Xu 2014; Yamashita et al. 2010). Some cathode-ray tube glass recyclers are reluctant to dispose of a cathode-ray tube and rather stockpiling cathode-ray tube glass due to the high cost of disposal and negative economic incentives. This stockpiling of cathode-ray tube possesses a great challenge to environmental pollution (soil and

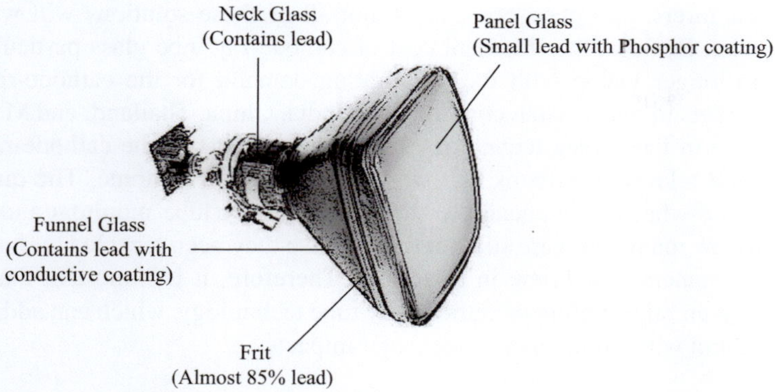

Fig. 1.1 Various components of cathode-ray tube. (Source: Author)

Fig. 1.2 Chemical component in cathode-ray tube glass structures in percentage weight. (Source: Adapted from Andreola et al. 2005; Chen et al. 2009; Shi et al. 2011; Singh et al. 2016a, b)

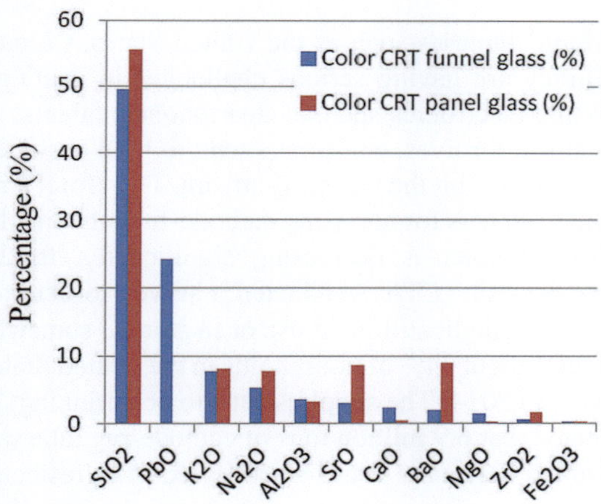

air) resulting from lead leaching. The detrimental environmental impact is the driving force to look for alternative technology and develop solutions for recycling cathode-ray tube glass.

Cathode-ray tubes are difficult to recycle because of the leaded glass they contain, and cathode-ray tube glass has essentially no commodity value. A typical cathode-ray tube has between 1.5 and 2 kg of lead (24.17 in percentage weight of total chemical components) among many other oxides as shown in Fig. 1.2. All of the lead is stored in the funnel component of the cathode-ray tube. The leaded glass needs to be processed, stored, and transported without causing environmental contamination. Recyclers typically must pay a consumer to take the panel and funnel glass. At the same time, the value of the remaining commodities, such as copper wiring, plastic, and other metals, has gone down in the past few years. That makes it much harder for recyclers to absorb the costs associated with cathode-ray tube recycling. Though there are other markets for the material, such as lead smelters, tile manufacturers, and glass companies, not all of these solutions will work for each recycler. Besides, the shipment cost of cathode-ray tube glass particularly in Asia is no longer viable with the evaporating demand for the cathode-ray tube in local markets in many Asian countries like India, China, Thailand, and Malaysia.

The surge in flat screen technology has replaced most of the cathode-ray tube monitors and television screens that we use at work and at home. The questions remaining are what has happened to those cathode-ray tube monitors and televisions, and how many more are still out there? Have they been recycled, or have they found a permanent new home in a landfill? Therefore, it is important that focus should be given on recycling of cathode-ray tube technology, which can address the above problem with minimal environmental impacts.

1.2 Current Status of the Waste Cathode-Ray Tube

Many countries such as the United States, China, India, Brazil, South Africa, and Turkey are having serious challenges in dealing with a waste cathode-ray tube. While electronics such as smartphones, tablets, and laptops seem to have shorter and shorter lives, consumers tend to hold on to televisions for more than a decade before they hit the recycling stream. Therefore, it is difficult to phase out completely the prospects for recycling cathode-ray tube in televisions in the near future though their number is decreasing significantly. In 2015, the Consumer Electronics Association (CTA) conducted a survey to determine how many cathode-ray tube devices might still be in use or in storage somewhere in US households. The study found about 34% of households in the United States still has at least one, down from 41% in 2014. The supply seems to be shrinking, but it is still substantial. It is estimated that 6.9 million tons of cathode-ray tube waste (43% of total E-waste) in the United States are yet to be collected from residential and commercial premises for recycling process (Singh et al. 2016a, b). The situation is further gloomy in Asian countries like China and India. Even countries like Turkey generated 0.22 kg cathode-ray tube per person in 2015 (Öztürk 2015). South Korea generated more than 8 million units of cathode-ray tube television waste in 2004–2005 out of which less than 3 million units, i.e., 37.5%, were recycled (Lee et al. 2007). It is estimated that 74 million televisions and 190 million personal computers will become obsolete in China (UNEP 2012), and cathode-ray tube monitor will comprise of 80% of E-waste (Song et al. 2012). It could be added that 6.3 million tons of waste was generated from cathode-ray tube monitor globally in 2014 (Baldé et al. 2015) with the majority of cathode-ray tube waste came from Asia as shown in Fig. 1.3.

With today's low scrap commodity prices, finding sufficient consumers of cathode-ray tube glass at a reasonable price is a significant challenge for the recycler. Failed cathode-ray tube recycling operations have made headlines in recent years because either the companies could not sustain financial backing or they could not find markets for the material. Some companies weighed down with cathode-ray tubes they could not recycle or move downstream and ended up stockpiling or

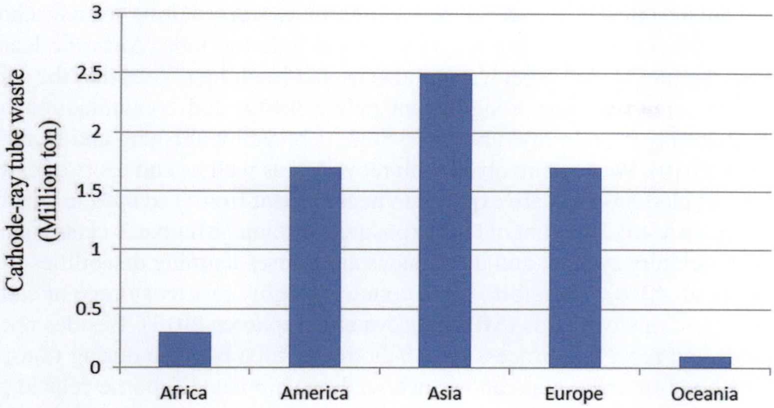

Fig. 1.3 Global cathode-ray tube waste resulting from the disposal of the monitor. (Source: Baldé et al. 2015)

dumping them. For example, when Phoenix opened in 2010, it promised to build furnaces in Ohio and Arizona to recycle cathode-ray tube glass into separate streams of glass and lead, and it received upfront financial backing to get the project going. Unfortunately, it found itself unable to meet processing demands and ended up with huge stockpiles of cathode-ray tube, much higher numbers than the Environmental Protection Agency (EPA) allows without a permit, which caused the company to cease operating.

The extended producer responsibility laws that promote electronics recycling also have complicated the process when it comes to a cathode-ray tube. Several electronics retailers had taken back programs that accepted cathode-ray tubes for free, in part because the companies only would get credit under extended producer responsibility (EPR) laws for electronics it collects for free. But once they introduced a recycling fee for cathode-ray tubes, volumes decreased. Disposal of the cathode-ray tube is a worldwide problem, generating concerns about widespread dumping not only within the country itself but also problem arising from transboundary shipment resulted from developed to developing countries. When cathode-ray tubes were still in use, numerous countries had recycling facilities, and vendors would ship to other countries to have the glass processed. With the last recycler and reuse company in India closing their doors to cathode-ray tube in 2015, the avenues for direct reuse of the glass are gone.

1.3 Disposal and Environmental Issues

There are basically two processes used for recycling of cathode-ray tube: (i) glass-to-glass recycling and (ii) secondary lead smelting. The interface between funnel and panel glass (frit line) is vulnerable to acidic environments and can cause potential environmental impacts on soil and groundwater resulting from leaching of lead in the funnel glass during recycling of cathode-ray tube. Airborne lead and particulate matter, leaded wash water and slag, and leaching of lead into the soil and groundwater could possess a significant public health and contamination of the stockyard during cathode-ray tube glass-to-glass recycling (Hsiang and Diaz 2011; Johri et al. 2010). Workers involved with recycling as well as end users of cathode-ray tube-recycled products are exposed to health threat from lead used in a cathode-ray tube. Even a small amount of lead exposure can damage human's central nervous system, circulatory system, and the kidneys and causes learning disabilities in children (Xu et al. 2016). Therefore, some countries highly restrict storage of cathode-ray tube waste in stockyards (Milovantseva and Saphores 2013). Besides not only stockyards and recycling process of a cathode-ray tube but also during transportation of cathode-ray tube glass and transboundary shipment, airborne-related pollution takes place. Air pollution also takes place during secondary lead smelters to extract lead by heating materials at high temperature. The emitted lead particles also contaminate soil and surface water as well as hazardous waste resulting from slag as a by-product.

1.4 Available Technologies for Recycling of Cathode-Ray Tube

There are various types of technology available for recycling of materials extracted from the cathode-ray tube. In general, the cathode-ray tube recycling process is categorized into two types: (i) open-loop and (ii) closed-loop recycling process as shown in Fig. 1.4.

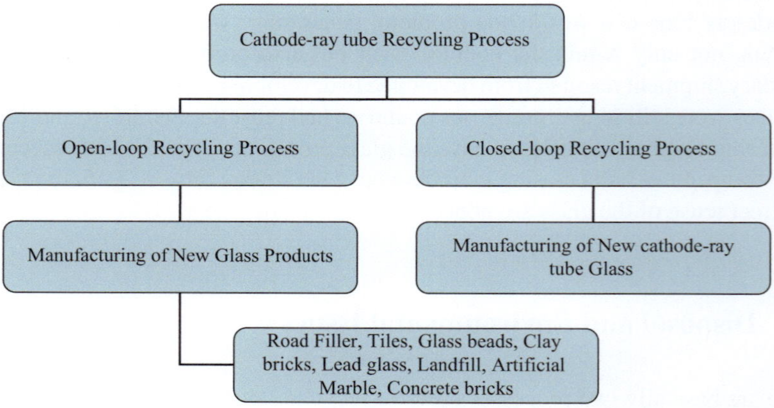

Fig. 1.4 An overview of cathode-ray tube recycling process. (Source: Adapted from Singh et al. 2016a)

Table 1.1 Open-loop and closed-loop recycling process

Open-loop process	Closed-loop process
Funnel and panel glass are segregated for use in different new products	Both funnel and panel glass are broken into small pieces without any segregation
New glass products such as tiles, road fillers, artificial marbles, glass beads, clay bricks, etc. are produced	Only new cathode-ray tube glass is produced
Attractive, feasible and economical, even to recycle funnel glass products particularly in ceramic industries	Many countries do not have manufacturers for cathode-ray tube glass products, therefore logistically and economically not feasible

1.4.1 Open-Loop Recycling Process

Open-loop recycling process (glass-to-lead recycling) uses recycled cathode-ray tube glass to manufacture new glass products. Obtaining raw cathode-ray tube glass during the open-loop process is not recommended because of various unknown compositions (Mueller et al. 2012). The applications of the raw cathode-ray tube in new products are considered as a barrier (Lee et al. 2012) as it requires the understanding of the composition of the raw cathode-ray tube (Lairaksa et al. 2013) unless the lead does not go beyond the standards (Yot and Mear 2011). Keeping in view this constraint of using funnel glass, open-loop process uses panel glass only for new glass products because of its constant and known composition (Iniaghe et al. 2013).

1.4.2 Closed-Loop Recycling Process

Closed-loop recycling process (glass-to-glass recycling) recycles old cathode-ray tube in the manufacturing of new cathode-ray tube. Both funnel and panel glass is broken into small pieces without any segregation in closed-loop process (Ertug and Unlu 2012). The closed-loop process was suitable when there was enough old cathode-ray tube available to meet the ongoing market demand (Lairaksa et al. 2013). The difference between the open- and closed-loop process is shown in Table 1.1.

1.5 Technology for Segregation of Funnel Glass from Panel Glass

The glass is one of the major components in a cathode-ray tube that requires special attention due to its various chemical composition during its recycling. Particularly, challenges arise when it comes to separation of funnel glass (lead) from panel glass (unleaded) of a cathode-ray tube. Therefore, recycling technology gives emphasis to avoid contamination and efficient extraction of lead. There are various recycling methods to segregate the funnel glass from panel glass.

1.5.1 Gravitational-Fall Method

Cathode-ray tube facing downward is dropped from a designated height due to the force of gravity resulting into the breaking of funnel glass by the impact of the surface. Thus, panel glass is separated from funnel glass during this method. The major challenge with this method is achieving clear separation. This method is low cost and simple (Herat 2008).

1.5.2 Electric-Wire Heating Method

In this method, thermal shock resulting from alternating heating and cooling is applied by wrapping a Nichrome along the frit line of the cathode-ray tube which results into breakage, thus separating the funnel glass from panel glass. It requires 1–3 min based on the size and type of cathode-ray tube (Lee et al. 2004). This method is suitable for mass production due to its easy operation, high efficiency, cost affordability, and absence of noise (Lee et al. 2004; Heart 2008). However, difficulty arises if the wire is improperly placed, the resulting creation of sharp edges on fragmented parts.

1.5.3 Thermal Shock

In this method, both funnel and panel glass along the frit line are segregated due to thermal shock resulting from restricted heat subsequently by cold air. This method uses electric wire heating and quench. This method is widely used in China.

1.5.4 Laser Cutting Method

It is similar to the thermal shock method, but in this method, the laser beam is used instead of electric wire heating followed by a spray with cold water (Yu et al. 2016). This energy-intensive method is suitable for wide-ranging processes (Heart 2008; Yu et al. 2016) although it is costly due to high investment. The drawbacks to this method are challenges of cutting thick glass and its restoration after the use of laser beam.

1.5.5 Diamond Cutting Method

This method can be carried out both under moist and parched conditions. During moist condition, diamond saw blades cut along the frit line of a cathode-ray tube with coolant water being sprayed while the cathode-ray tube kept in an enclosure keep rotating. The moist condition is suitable for all thickness and sizes of the cathode-ray tube. While during the parched process, diamond wheel and diamond belt with abrasive are used in the absence of water. The parched condition is more suitable for large-scale operation due to better performance in terms of clean separation (Lee et al. 2004; Heart 2008). However, the disadvantage for the application of parched condition is initially high investment costs. The diamond cutting method is faster than traditional mechanical cutting method due to its vacuum adsorption and dust recovery capacity, automatic edge searching, and laser positioning (Yu et al. 2016). This method is generally performed in three stages. In the first stage, a vacuum is removed and electron gun is dismantled. In the second stage, front panel, funnel glass, and shadow mask are all segregated, and in the last stage, front panel and funnel glass's coating is cleaned.

1.5.6 Water Jet Separation

High-pressured water containing abrasive materials in the form of the jet is sprayed through a nozzle along the frit line of the cathode-ray tube. This method is simple and affordable, but usually, the presence of abrasive materials contributes to wastewater which is a major concern (Heart 2008).

1.5.7 Acid Melting

Acid melting injects hot nitric acid and hot acid bath to dissolve the interface joining the panel and funnel of the cathode-ray tube. The major challenges in the use of this technology are the requirement for disposal of a large amount of wastewater and leachate generated during this method (Yu et al. 2016). Hence, this method is not suitable or recommended due to its adverse impact on the environment.

The above methods used for segregating funnel glass do not take into consideration in the presence of fluorescent colors in coatings. These coatings are of concern due to various metals which can be removed by the following technologies as shown in Table 1.2.

Table 1.2 Available technology for removal of coatings

Technology	Brief description	References
Wet-scrubbing method	Water and coatings scrubbed off from disintegrated cathode-ray tube glass positioned in a tumbling mill	Lee et al. (2004)
Ultrasonic method	Disintegrated cathode-ray tube glass is submerged into acid and water and immersed in an ultrasonic device for a certain time	Ezrat and Zhang (2014)
Sandblasting method	Air jet under high pressure ensures blasting of small steel balls onto the glass surface	Heart (2008)
Vacuum-suction method	Slack finishes from the panel glass are sucked using surface vacuum-suction device	Heart (2008)

1.6 Potential Technology for Removal of Lead from Cathode-Ray Tube

Lead has inimitable properties like resistance to corrosion, ductility, softness, and malleability and therefore is widely used in the manufacturing of batteries, solder, and X-ray shielding (Yu et al. 2016). It is quite difficult to extract a good amount of lead under normal pressure and temperature (Miyoshi et al. 2004) which comprises of O-Si-O- network and/or partly -O-Si-O-Pb-O- network encapsulated in the fissure of the glass assembly (Sasai et al. 2008). Therefore, to perform better lead extraction, importance is given on the selection of extraction lead methods by disintegrating the glass assembly. The well-known lead recovery process is as follows:

1.6.1 Pyrometallurgical Process

The pyrometallurgical process has been used successfully by various researchers to extracted lead from waste cathode-ray tube (Yot and Mear 2011; Okada and Yonezawa 2013, 2014; Mingfei et al. 2016). This process is used to remove lead and other metals by adding sodium carbonate powder (fusion agent), sodium sulfide (catalytic agent), and carbon powder as reducing agent with lead removal efficiency of 94% (Hu and Hui 2018) as shown in Fig. 1.5. Besides, Lu et al. (2013) used metallic iron for thermal reduction to extract lead with 99% efficiency from cathode-ray tube funnel glass. Addition of Na_2CO_3 also assists in the reduction process by restricting temperature below 1000 °C, to avoid lead evaporation (Okada and Yonezawa 2013). Recently Okada et al. (2015), using reducing and oxidizing condition, extracted lead from cathode-ray tube funnel glass dissolved into hydrochloric acid. Xing and Zhang (2011) used the pyrometallurgical process to extract nanoparticles of lead using a carbon-reducing agent, in a vacuum of 500–2000 Pa, with a temperature of 1000 °C for 2 h (Xing and Zhang 2011).

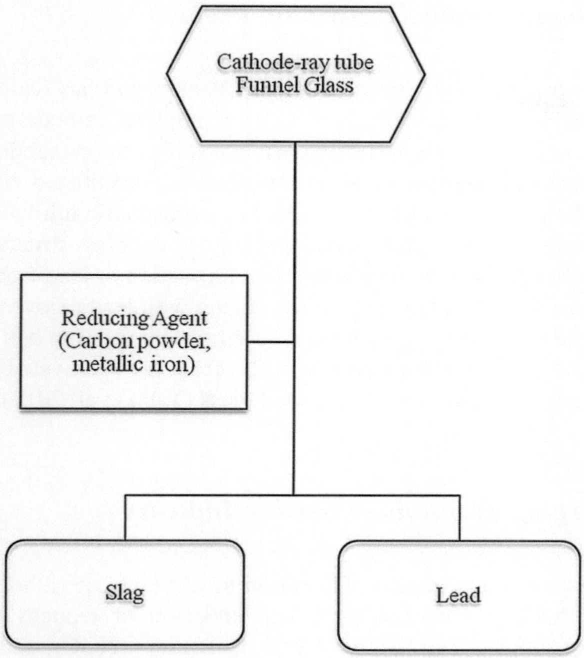

Fig. 1.5 Pyrometallurgical process. (Source: Adapted from Gong et al. 2016)

1.6.2 Hydrometallurgical Process

This method is used by applying acidic or caustic solutions on cathode-ray tube funnel glass followed by separation and cleansing procedures to concentrate the metals. However, to overcome the strong bonding of metals from glass and increase the solubility of lead, some pretreatments such as acid washing in 5% nitric acid for 3 h and rinsing with tap water on CRT funnel glass are carried out (Ling and Poon 2011). Zhang et al. (2013) used a strong alkaline solution combined with chemical leaching and mechanical activation (stirring ball mill) as a novel hydrometallurgical process with lead removal efficiency of 97%. Yuan et al. (2015) separated lead from cathode-ray tube funnel glass by pretreating with hydroxyl ions produced during an ion exchange reaction between metal ions in mechanically activated funnel glass and water under hydrothermal sulfidisation. Hydrometallurgy process is no longer profitable when transportation distance increases significantly.

1.6.3 Mechanochemical Activation Process

This process is triggered by physicochemical changes and has wide applications. Even without having the requirement of high temperature, cathode-ray tube funnel glass can be removed by using mechanochemical sulfidization reaction (Yuan et al. 2013). The mechanochemical sulfidization reaction is initiated by co-grinding cathode-ray tube glass with elemental sulfur in atmospheric nitrogen (Yuan et al. 2012). An extensive physical and chemical change of the glass structure is obtained through ball milling by mechanical activation. Extraction of lead from cathode-ray tube funnel glass is cost-effective and encouraging with lead recovery rate for funnel glass activated for 2 h at the rotational speed of 500 rpm (by ball mill) reached 92.5% (Yuan et al. 2012), compared with 1.2% from the inactivated sample (Yuan et al. 2013). It can also detoxify other leaded glass (Yaun et al. 2012).

1.6.4 Emerging Technology Used in Industry

Industries are also paying increased attention to recycling of cathode-ray tube by introducing technology that results in less emission or reduces environmental impacts. For example, The Nulife Glass Company (NGC) and the Sweeep Kuusakoski Company (SKC) in the United Kingdom (UK) use chemicals and furnace for removal of lead from waste cathode-ray tube in a sustainable process without any residual waste. This recycling furnace technology can process funnel glass 10 tons per day equivalent to around 60 tons cathode-ray tube televisions throughout its life cycle (Nulife Glass 2015). Sweeep Kuusakoski Company in its recycling process segregated funnel and panel glass without any cross-contamination of the two different glass types as shown in Fig. 1.6. An electrolytic converter crushes the funnel glass and produces very distinct and clean molten glass and lead without having any waste left. Cathode-ray tube recycling furnace further adds values to the production process by removal of 1 kg of lead from unused or left-out glass (Sweeep kuusakoski 2015). The introduction of this new emerging technology adopted in the above industries initiates a new dimension in the recycling of cathode-ray tube waste. The technology is highly energy efficient which uses $0.50 electricity and saves $2 worth of lead and clean glass for each treatment of television, and requires no expensive extraction or filtration system due to its very negligible emissions (Sweeep Kuusakoski 2015).

1.6.5 Construction Materials

Many researchers have used waste cathode-ray tube funnel glass for the purpose of making glass-ceramic brick and concrete (construction materials) due to additive properties of funnel glass. Dondi et al. (2009) proposed assorted cathode-ray tube

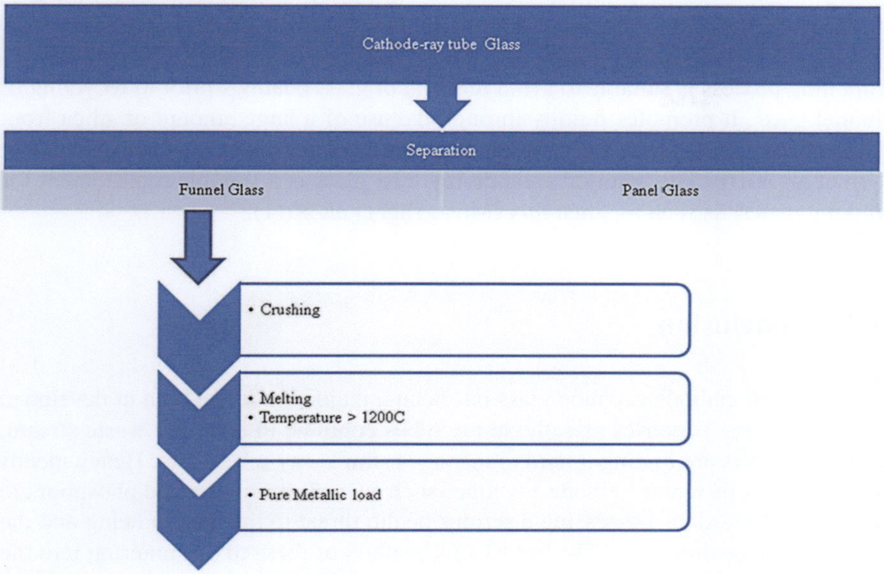

Fig. 1.6 Cathode-ray tube glass recycling process. (Source: Adapted from Singh et al. 2016b)

funnel and panel glass of particle size below 1 mm to replace silica for production of clay bricks and roof tiles as construction materials. The study conducted by Dondi et al. (2009) showed that bricks and roof tiles quality was good enough. Finely ground glass has the cementitious ability, and it forms calcium silicate hydrate responsible for the development of high strength in concrete when its silica content reacts with calcium hydroxide in the presence of water (Shi et al. 2005). Hence, cathode-ray tube funnel glass is used to make concrete blocks with acceptable compressive strength (Ling and Poon 2014). These concrete blocks have lower drying shrinkage, i.e., more resistant to water with high block density, making the blocks suitable against gamma radiation.

1.6.6 Waste Vitrification Glass

The technology is reliable and converts cathode-ray tube waste into glass-forming materials in high-temperature melters through very fine grinding and melting (Engelhardt 2013). The funnel glass can be used for nuclear waste vitrification. The waste comprising hazardous and nuclear substances are confined in crystalline structure, while the lead offers necessary protection against radiation (gamma) through shielding. Trombay nuclear plant in India uses lead oxide (PbO) with vitrification glass to stabilize nuclear waste (Sengupta et al. 2013).

1.6.7 Smelting Flux

Smelting process is suitable to avoid removal of glass coatings prior to recycling of funnel glass. It promotes fluidity through the use of a large amount of silica from lead and copper smelter, thus extract and convey impurities to the processed slag (Yu et al. 2016). The assorted cathode-ray tube glass is a suitable replacement for fluxing materials such as silica flux (Mostaghel et al. 2011).

1.7 Conclusion

Recycling of cathode-ray tube glass has been a major challenge both at developed and developing countries as cathode-ray tubes continue to enter the waste stream, while recyclers are finding it hard to manage cathode-ray tube waste. Heavy metals and other toxins within cathode-ray tubes such as lead, strontium, and phosphor are extremely hazardous possessing a serious health threat to the human being and the surrounding environment. The most likely method of these toxins entering into the human system is by leaching from landfills into the soil and groundwater. Research into new and more effective methods of recycling is needed to handle the volume of E-waste that is created on a daily basis. Older technology requires special attention because as the volume of the devices going into the recycling industry falls, the financial viability of the recycling effort drops faster than the number of devices left. We need processes that work on specific problems like the separation of glass and lead into two sellable components that make a positive impact on toxic global problems. The emerging technology using a furnace and chemicals for extraction of toxic lead from the cathode-ray tube in an environmentally sustainable way without any residual waste as introduced by the Nulife Glass Company and the Sweeep Kuusakoski Company in the United Kingdom looks promising. However, a combination of regulatory enforcement, better funding mechanisms, and creative problem-solving could be the key to moving cathode-ray tube recycling forward in a healthy way. Besides, finding good solutions for cathode-ray tube glass recycling will help set a precedent for the next end-of-life electronics that have their own challenges, such as used electronics that contain mercury or other hazardous substances.

References

Andreola F, Barbieri L, Corradi A, Lancellotti I (2005) Cathode ray tube glass recycling: an example of clean technology. Waste Manag Res 23:314–321. https://doi.org/10.1177/0734242X05054422

Baldé CP, Wang F, Kuehr R, Huisman J (2015) The global e-waste monitor-2014. United Nations University, IAS–SCYCLE, Bonn, Germany. Available at https://i.unu.edu/media/unu.edu/news/52624/UNU-1stGlobal-E-Waste-Monitor-2014-small.pdf

Chen M, Zhang FS, Zhu JX (2009) Detoxification of cathode ray tube glass by the self-propagating process. J Hazard Mater 165:980–986. https://doi.org/10.1016/j.jhazmat.2008.04.084
Dondi M, Guarini G, Raimondo M, Zanelli C (2009) Recycling PC and TV waste glass in clay bricks and roof tiles. Waste Manag 29:1945–1951. https://doi.org/10.1016/j.wasman.2008.12.003
Engelhardt T (2013) Vitrification of nuclear waste, the winning proposal of the CRT challenge of the Consumer Technology Association, Digital Dialogue, 2013. Available at http://www.ce.org/Blog/Articles/2013/October/CRT-Challenge-The-Winning-Proposal.aspx
Ertug B, Unlu N (2012) An evaluation study: recent developments and processing of glass scrap recycling. Epd Congress:381–388. https://doi.org/10.1002/9781118359341
Erzat A, Zhang F-S (2014) Detoxification effect of chlorination procedure on the waste lead glass. J Mat Cycling Waste Manag 16(4):623–628. https://doi.org/10.1007/s10163-014-0290-6
Gong Y, Tian XM, Wu YF, Zhe T, Lei L (2016) Recent development of recycling lead from scrap CRTs: a technological review. Waste Manag 57:176–186. https://doi.org/10.1016/j.wasman.2015.09.004
He Y, Xu Z (2014) The status and development of treatment techniques of typical waste electrical and electronic equipment in China: a review. Waste Manag Res 32(4):254–269. https://doi.org/10.1007/s11356-018-2136-6
Heart S (2008) Recycling of cathode ray tubes (CRTs) in electronic waste. Clean Soil Air Wat 36:19–24. https://doi.org/10.1002/clen.200700082
Hsiang J, Díaz E (2011) Lead and developmental neurotoxicity of the central nervous system. Curr Neurobiol 2(1):35–42. https://pdfs.semanticscholar.org/ed4e/57f3e2f3c3f7231af00427ab5271907cdfa0.pdf
Hu B, Hui W (2018) Lead recovery from waste CRT funnel glass by high-temperature melting process. J Hazard Mater 343(5):220–226. https://doi.org/10.1016/j.jhazmat.2017.09.034
Iniaghe PO, Adie GU, Osibanjo O (2013) Metal levels in computer monitor components discarded in the vicinities of electronic workshops. Toxicol Environ Chem 95:1108–1115. https://doi.org/10.1080/02772248.2013.863890
Johri N, Grégory J, Robert U (2010) Heavy metal poisoning: the effects of cadmium on the kidney. Biometals 23:783–792. https://doi.org/10.1007/s10534-010-9328-y
Lairaksa N, Moon AR, Makul N (2013) Utilization of cathode ray tube waste: encapsulation of PbO-containing funnel glass in Portland cement clinker. J Environ Manag 117:180–186. https://doi.org/10.1016/j.jenvman.2012.12.014
Lee CH, Chang CT, Fan KS, Chang TC (2004) An overview of recycling and treatment of scrap computers. J Hazard Mater 114:93–100. https://doi.org/10.1016/j.jhazmat.2004.07.013
Lee C-H, Chang S-L, Wang K-M, Wen L-C (2007) Present status of the recycling of waste electrical and electronic equipment in Korea. Resour Conserv Recycl 50:380–397. https://doi.org/10.1016/j.resconrec.2007.01.010
Lee J-S, Cho S-J, Han B-H, Seo Y-C, Kim B-S, Heo SP (2012) Recycling of TV CRT panel glass by incorporating to cement and clay bricks as aggregates. Adv Biomed Eng 7:257
Ling TC, Poon CS (2011) Utilization of recycled glass derived from cathode ray tube glass as fine aggregate in cement mortar. J Hazard Mater 192:451–456. https://doi.org/10.1016/j.jhazmat.2011.05.019
Ling TC, Poon CS (2014) Use of CRT funnel glass in concrete blocks prepared with different aggregate-to-cement ratios. Green Mater 2(1):43–51. https://doi.org/10.1680/gmat.13.00013
Lu X, Shih K, Liu C, Wang F (2013) Extraction of metallic Lead from Cathode ray tube (CRT) funnel glass by thermal reduction with metallic Iron. Environ Sci Technol 47(17):9972–9978. https://doi.org/10.1021/es401674d
Milovantseva N, Saphores JD (2013) E-waste bans and U.S. households' preferences for disposing of their e-waste. J Environ Manag 124:8–16. https://doi.org/10.1016/j.jenvman.2013.03.019
Mingfei X, Yaping W, Jun L, Hua X (2016) Lead recovery and glass microspheres synthesis from waste CRT funnel glasses through carbon thermal reduction enhanced acid leaching process. J Hazard Mater 305:51–58. https://doi.org/10.1016/j.jhazmat.2015.11.032
Miyoshi H, Chen D, Akai T (2004) A novel process utilizing subcritical water to remove lead from wasted lead silicate glass. Chem Lett 33:956–957. https://doi.org/10.1246/cl.2004.956

Mostaghel S, Yang Q, Samuelsson C (2011) Recycling of cathode ray tube in metallurgical processes: influence on environmental properties of the slag. Glob J Environ Sci Technol 1:19

Mueller JR, Boehm MW, Drummond C (2012) Direction of CRT waste glass processing: electronics recycling industry communication. Waste Manag 32:1560–1565. https://doi.org/10.1016/j.wasman.2012.03.004

Nulife Glass (2015) Recycling CRTs from televisions & computer screens. Available at: http://www.nulifeglass.com/

Okada T, Yonezawa S (2013) Energy-efficient modification of reduction-melting for lead recovery from cathode ray tube funnel glass. Waste Manag 33:1758–1763. https://doi.org/10.1016/j.wasman.2013.04.009

Okada T, Yonezawa S (2014) Reduction-melting combined with a Na2CO3 flux recycling process for lead recovery from cathode ray tube funnel glass. Waste Manag 34:1470–1479. https://doi.org/10.1016/j.wasman.2014.04.012

Okada T, Nishimura F, Yonezawa S (2015) Removal of lead from cathode ray tube funnel glass by combined thermal treatment and leaching processes. Waste Manag 45:343–350. https://doi.org/10.1016/j.wasman.2015.05.008

Öztürk T (2015) Generation and management of electrical–electronic waste (e-waste) in Turkey. J Mat Cycles Waste Manag 17:411–421. https://doi.org/10.1007/s10163-014-0258-6

Sasai R, Kubo H, Kamiya M, Itoh H (2008) Development of an eco-friendly material recycling process for spent lead glass using a mechanochemical process and Na2EDTA reagent. Environ Sci Technol 42:4159–4164. https://doi.org/10.1021/es0719576

Sengupta P, Kaushik C, Dey G (2013) Immobilization of high level nuclear wastes: the Indian scenario, on a sustainable future of the Earth's natural resources. Springer, Berlin/Heidelberg, pp 25–51. https://doi.org/10.1007/978-3-642-32917-3_2

Shi C, Wu Y, Riefler C, Wang H (2005) Characteristics and pozzolanic reactivity of glass powders. Cem Concr Res 35:987–993. https://doi.org/10.1617/s11527-012-9880-8

Shi X, Li G, Xu Q, He W, Liang H (2011) Research progress on recycling technology of end-of-life CRT glass. Mater Rev 11:129–132

Singh N, Wang J, Li J (2016a) Waste cathode rays tube: an assessment of global demand for processing. Procedia Environ Sci 31:465–474. https://doi.org/10.1016/j.proenv.2016.02.050

Singh N, Li J, Zeng X (2016b) Solutions and challenges in recycling waste cathode-ray tubes. J Clean Prod 133:188–200. https://doi.org/10.1016/j.jclepro.2016.04.132

Song Q, Wang Z, Li J, Zeng X (2012) Life cycle assessment of TV sets in China: a case study of the impacts of CRT monitors. Waste Manag 32:1926–1936. https://doi.org/10.1016/j.wasman.2012.05.007

Sweeep Kuusakoski (2015) Available at: http://www.sweeepkuusakoski.co.uk/

UNEP (2012) Illicit trade in electrical and electronic waste (e-waste) from the world to the region, chapter 9, pp 101–111. Available at https://www.unodc.org/documents/toc/Reports/TOCTA-EA-Pacific/TOCTA_EAP_c09.pdf

Xing MF, Zhang FS (2011) Nano-lead particle synthesis from waste cathode ray-tube funnel glass. J Hazard Mater 194:407–413. https://doi.org/10.1016/j.jhazmat.2011.08.003

Xu Q, Yu M, Kendall A, He W, Li G, Schoenung JM (2016) Environmental and economic evaluation of cathode ray tube (CRT) funnel glass waste management options in the United States. In: Kirchain RE et al (eds) REWAS 2016. Springer, Cham. https://doi.org/10.1007/978-3-319-48768-7_48

Yamashita M, Wannagon A, Matsumoto S, Akai T, Sugita H, Imoto Y, Komai T, Sakanakura H (2010) Leaching behavior of CRT funnel glass. J Hazard Mater 184:58–64. https://doi.org/10.1016/j.jhazmat.2010.08.002

Yot PG, Mear FO (2011) Characterization of lead, barium and strontium leach- ability from foam glasses elaborated using waste cathode ray-tube glasses. J Hazard Mater 185:236–241. https://doi.org/10.1016/j.jhazmat.2010.09.023

Yu M, Liua L, Li J (2016) An overall solution to cathode-ray tube (CRT) glass recycling. Procedia Environ Sci 31:887–896. https://doi.org/10.1016/j.proenv.2016.02.106

Yuan W, Li J, Zhang Q, Saito F (2012) Innovated application of mechanical activation to separate lead from scrap cathode ray tube funnel glass. Environ Sci 46(7):4109-14. https://doi.org/10.1021/es204387a

Yuan W, Li J, Zhang Q, Saito F, Yang B (2013) Lead recovery from cathode ray tube funnel glass with mechanical activation. J Air Waste Manage Assoc 63(1):2–10. https://doi.org/10.1080/10962247.2012.711796

Yuan W, Meng W, Li J, Zhang C, Song Q, Bai J, Wang J, Li Y (2015) Lead recovery from scrap cathode ray tube funnel glass by hydrothermal sulphidisation. Waste Manag Res:1–7. https://doi.org/10.1177/0734242X15597777

Zhang C, Wang J, Bai J, Guan J, Wu W, Guo C (2013) Recovering lead from cathode ray tube funnel glass by mechano-chemical extraction in alkaline solution. Waste Manag Res 31:759–763. https://doi.org/10.1177/0734242X13484188

Chapter 2
Reconfigurable Recycling Systems of E-waste

Piotr Nowakowski

Contents

Abstract E-waste collected from households must be properly processed in recycling plants to acquire high-purity output materials and hazardous substances removed. It requires a systematic approach to the disassembly process and wastes classification and categorization. Big variety of shapes and materials used in the equipment need processing lines including manual and automated sections. The main purpose of the machines used on the processing lines is to shred equipment into small size fraction and then separate each material depending on the physical properties. In such case, the output material from the E-waste disassembling plants can be recycled and used in new parts or components. This chapter includes a description of methods of disassembly focused on E-waste recycling in compliance with environmental standards. The required steps of the end-of-life products disassembly vary depending on the category of waste equipment. To show these differences, the chapter includes two case studies showing the configuration of a layout of E-waste processing lines with possible options to reconfigure them. The variants of the system's configuration depend on the volume of the waste stream, labor cost, and required purity of output materials. Economic efficiency indicator of E-waste processing indicates big differences in potential profit from recycling E-waste

P. Nowakowski (✉)
Faculty of Transport, Silesian University of Technology, Katowice, Poland
e-mail: Piotr.Nowakowski@polsl.pl

A. Khan et al. (eds.), *E-waste Recycling and Management*,
Environmental Chemistry for a Sustainable World 33,
https://doi.org/10.1007/978-3-030-14184-4_2

mainly depending on labor cost. Example of calculation of this indicator has been presented in this chapter on cooling appliances recycling for four European countries.

Keywords E-waste · Recycling system · Waste processing · Processing line configuration · Cost and revenues of recycling · Electrical and electronic equipment · The efficiency of recycling · Disassembly · Modular system · Shredding and sorting

2.1 Introduction

A big variety of electrical and electronic products are commonly used worldwide. These products at the end of life become E-waste or waste electrical and electronic equipment (European Union 2003). They have to be properly collected and recycled. It depends on regulations in individual countries. One of the most important issues with recycling E-waste is concerned with environmental impact hazardous substances included in the equipment (Barba-Gutiérrez et al. 2008) and high recycling potential of each piece of waste. In many countries, waste electrical and electronic equipment management systems have been adopted (Chung and Zhang 2011; European Commission 2012; Dwivedy and Mittal 2012). In some countries, all categories of E-waste must be collected. It is an obligation imposed by the European Union directive or other country legislations. The main reason is the protection of the natural environment and human health against hazardous substances. At the same time, all E-waste can be legally collected and transferred to a recycling plant where the materials can be recovered. Other countries have introduced selective categories of waste electrical and electronic equipment with an obligation of a collection in the USA (Horner and Gertsakis 2006; Kahhat et al. 2008), Switzerland (Hischier et al. 2005), Taiwan, and Japan (Lee et al. 2007; Chung and Murakami-Suzuki 2008). These wastes are characterized by a large diversity of material composition and a different approach to disassembly and recycling (Oguchi et al. 2011). Basic issues considering recycling of E-waste should focus on the volume of the waste stream, environmental requirements imposed by legislation, recycling plant performance, operating costs – especially labor cost – and the ability to process further fraction.

Figure 2.1 presents the most important issues that have to be taken into consideration when disassembling and recycling waste equipment. The legislation creates the framework of the system operation and imposes requirements for residents, collection companies, and recycling plants. From environment point of view, it is necessary to include all the impacts on the natural environment from including potential hazard of contamination but also a chance to recycle materials used in these products (Ikhlayel 2017). At the recycling stage, the primary goal is to disassemble waste equipment in compliance with legislative requirements and standards (ACRR 2009; CENELEC 2013). Therefore, the main activity at the initial stage focuses on the

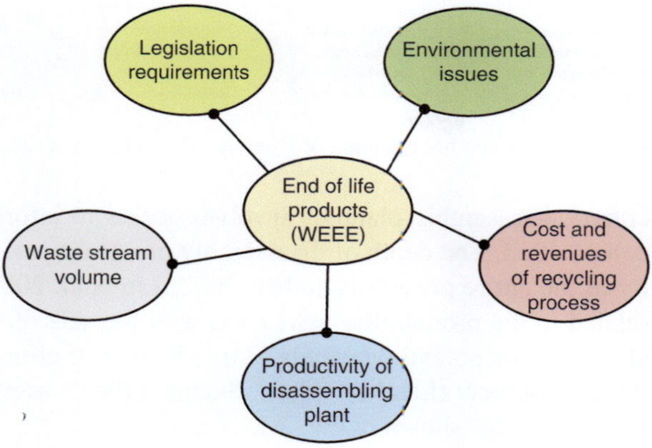

Fig. 2.1 The most important issues to be analyzed from the E-waste recycling system perspective

removal of hazardous substances from the waste equipment and then sorting materials to be processed (European Commission 2012). A design of a disassembly plant must take into consideration the most waste stream of waste electrical and electronic equipment generated in the area covered by collection and should include estimated volume from households and other non-household sources (Capraz et al. 2015).

The productivity of a recycling plant depends on the disassembling strategy. Manual disassembly is less productive, but many elements and assemblies can be easily identified by workers conducting this process (Chung and Peng 2005; Capraz et al. 2017). Preselecting of the components can acquire homogeneous material components or valuable parts. Example of such an approach can be a removal of the counterweights from washing machines. These are concrete parts and after taking out from a washing machine, they can be stored separately. A transfer of these components into a shredder could possibly affect its operation and breaking. In a similar way, disassembling a processor from the computer's mainboard and separating it is more profitable than shredding complete mainboards, including processors. The waste stream volume should be adjusted to a recycling plant capacity. Planning the waste processing lines in a plant must take into consideration, the scrap material prices depend on global markets, and the labor cost has the main impact on a potential profit from E-waste recycling.

2.2 Waste Electrical and Electronic Equipment Disassembly Planning

Planning the disassembling of waste electrical and electronic equipment is a demanding task. Such equipment has various shapes, masses, material compositions, and techniques of joining elements. Therefore, it is difficult to automate completely this process. The best case occurs when the equipment belongs to the same category and group, e.g., refrigerators, television sets, washing machines, etc. The

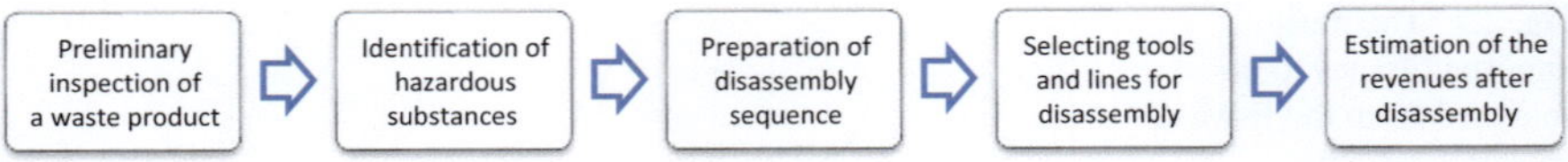

Fig. 2.2 Disassembly planning of E-waste. The main steps to be taken in a recycling plant

correct disassembly planning involves access to information on the parts material composition. The depth of disassembly must take into consideration the profitability of the entire process (Lambert 2002; Liu et al. 2009). This criterion is not only related to the profitability associated with the sale of the recycled materials, but it also includes potential benefits of a reduction or elimination of harmful effects on the environment (Lambert 2003). Planning the disassembly process should include five basic steps shown in Fig. 2.2.

Preliminary inspection of waste products begins before the main dismantling sequence. Its purpose is to specify a product material inventory and joints of the assemblies. To find the optimal disassembly path, consider the following issues (Li et al. 2006):

- A potential value of components to be dismantled
- Identification of components
- Choosing a disassembly strategy

When identifying components, some of them can be dismantled as a whole, if the entire subassembly has more value on the secondary market and can be sold to another plant. It is crucial in planning a dismantling sequence of the end-of-life product. For example, when the ultimate purpose of disassembly is the recovery of homogeneous materials, collecting all components made of steel, plastic, or glass prevents unnecessary dismantling steps (Lee et al. 2001). The choice of dismantling strategy depends on the economic analysis of the process. The aim is to compare the profitability rate of materials and parts recovery in relation to alternative solutions.

The following issues can be taken into consideration to generate the disassembly sequence (Hsin-Hao (Tom) Huang et al. 2000; Bogaert et al. 2008; Duta et al. 2008; Kuo 2010; Nowakowski 2018):

- Fulfilling the recommendations of waste electrical and electronic equipment directives or regulations specific to an individual country; the purpose is to remove elements that are hazardous to the environment, such as batteries, printed circuit boards, liquid crystal displays, refrigerants, electrolytic capacitors, etc.
- Manual disassembly for separating the main components to achieve maximum profit with a short disassembly time; targeting mainly on ferrous, nonferrous, and precious metals, e.g., gold in processors, mainboards, gold-plated contacts, etc.
- Adapted to the existing installation in a large disassembling and recycling plant with automated shredding, separating, and sorting of materials.
- Using modern identification methods such as barcodes, two-dimensional codes, or radio-frequency identification technology, with the possibility of automating various stages of disassembly.

Selection of disassembly tools is a necessary step to properly carry out the dismantling sequence and prepare a team of workers. Disassembly can be carried out manually, on semiautomated or automated lines and using systems equipped with robots (Kang and Schoenung 2005; Williams 2006; Li et al. 2017). Selection of tools is easier for larger batches of waste equipment belonging to the same category.

Economic evaluation of the disassembly process must take the following factors into consideration:

- Disassembly time
- Disassembly costs – including labor cost, transportation and processing of materials, tools, machines, and total power consumption
- Cost of the utilization of hazardous substances
- Revenues – based on the market price of acquired materials and components to be sold to other enterprises

Table 2.1 presents the main attributes of possible disassembly methods including manual, semiautomated, automated, and supported by robots (Santochi et al. 2002; ElSayed et al. 2012).

2.3 E-waste Processing Costs and Revenues

Each disassembly method incurs fixed and variable costs. For installations capable of processing a high volume of waste electrical and electronic equipment and where labor cost is high, the only rational solution is a combination of automated or robotic line with the minimal necessary support of manual disassembly (Kang and Schoenung 2006; Bakar and Rahimifard 2008). In the countries where the labor cost is low, manual dismantling methods are commonly applied (Wath et al. 2011; Li et al. 2013). After disassembly, each recycling plant can offer various types of materials for sale for third parties. Their purity depends on the applied technology in sorting. As the output, it can be ferrous and nonferrous metals, plastics of various grain sizes and also complete parts, or assemblies (e.g., compressors, printed circuit boards – mainboards, memory cards, etc.). Large parts or assemblies of homogenous materials can be removed and separated manually – their potential value can be higher than processed material acquired after shredding and separation.

Figure 2.3 shows the components – copper pipes removed from refrigerators after manual cutoff and fine grains of copper after shredding and separating, received at the end. The processed fraction of copper is high purity but consists of other metals, and therefore the price offered by copper smelter is lower than 10–20% of the manually disassembled copper components.

Separate lines for disassembling various categories of waste electrical and electronic equipment should be installed in a recycling plant. Each of them should be capable of providing specific tasks for the removal of hazardous substances, disassembling individual components, including power supply units, printed circuit

Table 2.1 Characteristics of the types of E-waste disassembly methods

Type of disassembly	Description	Machines/ tools	Productivity	Focus on
Manual	Preparation for other stages of dismantling. Obligatory for some categories of waste. Removal of substances in compliance with legislation	Simple tools	Low but efficient for assemblies containing precious or rare earth metals	An opening of equipment casing removal of homogenous material parts or others of high value and also hazardous substances
Semiautomated	Using power tools for faster removal of heavy components or to access valuable parts for easier disassembly	Hydraulic, pneumatic, electric tools	Medium, supporting manual method. It allows executing quick removal of heavy components	Removal of heavy objects or assemblies from large home appliances
Automated	Alignment of machines with different configurations to obtain homogenous fractions of materials, automated transport, and handling. Using conveyors and vibrating feeders	Shredding, grinding, milling machines, automated separation	High, the process is on the specialized machines, transportation of the materials, semi-processed, or final is by means of conveyors or feeders	Mechanical processing to obtain homogenous small size grains of materials for easy separation and sorting
Robotic	Designed for equipment of similar construction, mostly belonging to the same group from the category, TV sets, refrigerators, mobile phones	Robots, sensor devices	Efficient for similar products, a potential for future development	Full automation of disassembly, possible replacement of human work, safe handling of hazardous substances

boards, cables, and heavy components. Disassembly of cooling appliances or air-conditioning units must begin with the removal of refrigerants (Keri 2012). Removal of hazardous substances from the waste equipment brings improvement of the output material purity. Any contamination can be a reason for lowering the price of the entire shipment of the processed material.

The information technologies equipment category requires different approach disassembling. Printed circuit boards used in this equipment often contain copper and precious metals including gold and silver. Some parts or assemblies include also rare earth metals. The components and assemblies from personal computers and notebooks have to be classified into groups. The highest market values can be reached by processors, mainboards, and memories (Hall and Williams 2007; Cucchiella et al. 2016; Rosa and Terzi 2016). A more challenging task is disassembly and recycling of mobile phones and smartphones (Tanskanen 2012; Sarath et al.

Fig. 2.3 Copper components from manual disassembly (left) and copper fraction from the processing line. This material comes from refrigerators and cooling appliances

2015). These items consist of casings, screens, touch screens, batteries, and mainboards. The older generation of mobile phones was easier to disassemble. A battery removal was simple. Currently, the batteries are integrated into the devices and manual removal requires a long time. In such a case, the labor cost exceeds potential revenues (Terazono et al. 2015).

Estimation of economic efficiency indicator W_p^t for a disassembly and recycling line in a plant can be calculated using the following formula – 1 (Nowakowski 2017):

$$W_p^t = \frac{\sum_{i=1}^{d} \sum_{j=1}^{m(i)} r_{i,j}}{\sum_{i=1}^{d} \sum_{j=1}^{m(i)} c_{i,j}} \tag{2.1}$$

where W_p^t, the efficiency indicator of processing waste equipment using t technology; t, a number of the technology in the recycling plant; $r_{i,j}$, revenues after sale of *j-th* material to third parties [euro]; $m(i)$, a number of raw materials acquired after processing i-th piece of waste; d, a number of waste equipment for processing; and $c_{i,j}$, total cost of disassembly and processing (labor, energy, depreciation, and others) of i-th the piece of waste for j-th material using t technology.

And total efficiency indicator $W_p$for all disassembly lines in a plant can be calculated by formula – 2:

$$W_p = \sum_{i=1}^{\bar{T}} \alpha_t \cdot W_p^t \tag{2.2}$$

where $T = \{t : 1,2,\ldots,\bar{T}\}$, set of disassembly technology numbers, and $\bar{T}$, a number of disassembling technologies.

Table 2.2 Main parameters of the cooling appliances processing line

Parameter	Unit	Value
Power consumption	[kW]	310
Capacity	[pcs/h]	30
Required personnel	[persons]	6

Table 2.3 A potential revenue from refrigerators recycling per 1 h operation of recycling line

Material	Average material content in processed appliance [%]	Mass [kg/h]	Scrap price [EUR/t]	Revenue [EUR/h]
Ferrous metals	57	855	280	239
Plastics PS	27	405	270	109
Aluminum	5.7	85.5	1650	141
Copper	6	90	5340	480
Total revenue [EUR/h]				**969**

The output material from the processing line depends on sorting and separation methods applied. It can be transported to other plants – aluminum and copper smelters, steel works, or others. The income from sales depends on the material type and volume of the categories of equipment being processed.

The exemplary calculations have been conducted for the cooling equipment processing line. The main parameters of the disassembly line are shown in Table 2.2 (Vary 2018).

Table 2.3 shows the main output materials acquired from the cooling equipment. Material inventory is limited to ferrous metals, copper, aluminum, and plastics (polystyrene) (Huisman and Magalini 2007). All parameters adopted for the calculation and shown in Tables 2.3 and 2.4 concern the operation of the installation for 1 h and include average scrap prices for March 2018 (GRN 2018; LME 2018). The price of the entire line is 500.000 euro.

After substituting revenues and operational parameters, we obtain an efficiency indicator calculated for four countries. Commodity prices, including scrap metals, are set on global stock exchanges; therefore they are assumed constant for each country. The main difference is the labor (BMAS 2018; GUS 2018; INSEE 2018; SWI 2018) and electricity cost (Eurostat 2018; SWISSGRID 2018). As a result, the efficiency indicator W_p is much higher for Poland and Romania compared to Germany or Switzerland (Table 2.4).

2.4 Case Study: Example E-waste Recycling Plant Layout for Different Categories of Equipment

A recycling plant for processing various categories of E-waste should be equipped with several disassembly lines and specialized machines. Inclined and trough belt conveyors will be used for transportation of material between individual processing

Table 2.4 The cooling equipment processing efficiency factor for different countries (calculated for 1 h of operation)

	Costs [EUR]				Revenues [EUR]				W_p
Country	Labor	Energy	Depreciation	Others	Ferrous Metals	Cooper	Aluminum	Plastics PS	
Germany	52.8	47.12	15	10	239	480	141	109	**7.8**
Poland	21.6	27.28	15	10	239	480	141	109	**13.1**
Romania	15	23.87	15	10	239	480	141	109	**15.2**
Switzerland	126	45.26	15	10	239	480	141	109	**4.94**

machines. Vibratory feeders will distribute material onto transferring conveyor from the outlet of shredders or hammer mills. An automated recycling line should be equipped with the following separators:

- Ferrous metals separator
- Eddy current separator
- Gravity separator
- Screening separator

The supply of the waste electrical and electronic equipment should be constant; otherwise, such plant could bring losses. A case study presented in this chapter is designed to process E-waste stream from one million households. On the average, the mass of waste equipment collected from one household is 30 kg/year. To calculate this value, the expected value of the life of each product has been included (Cooper 2004; Nowakowski 2016). The plant productivity is set to 30.000 [tonnes/year] for two shifts for 300 days a year, and it is sufficient to process equipment from one million households.

Considering different categories of E-waste and methods of disassembly, the plant should be equipped with the following installations to achieve the assumed processing efficiency:

- The line for processing cooling and air-conditioning equipment with a capacity of 35–40 units/h
- The line of disassembling flat television sets and monitors – 50 items/h
- The line for the processing of small equipment together with the casings and components of large home appliances after manual dismantling
- Manual dismantling station equipped with the necessary tools
- A machine for cable recycling

The required number of employees and the total mass of E-waste belonging to an individual category are shown in Table 2.5. It includes estimated waste stream from households, disassembly, and handling time for each category of waste (Nowakowski 2015).

Figure 2.4 shows manual disassembly section of the plant. It is fully configurable and other stations can be set up additionally. After manual dismantling section waste equipment should be transferred to an automated processing system. The main purpose is to prepare waste equipment to be transferred to the automated section of the processing system.

E-waste delivered to the plant would be subjected to initial disassembly carried out manually using basic tools. It will take place on specially adapted steel tables of high resistance to damage and loads. The stands will be equipped with pneumatic and electric tools needed for disassembly, as well as cutters, pliers, etc. During the initial dismantling of waste equipment, elements containing toxic or hazardous substances (e.g., batteries, mercury switches, electrolyte capacitors) should be removed. Such components have to be stored in separate containers, and then they will be transferred to plants specializing in their disposal.

Table 2.5 Processing capacity and employees required for different categories of E-waste

Category of WEEE	Number of employees per shift	Total annual mass of waste for processing for two shifts and 300 working days [t]
Cooling equipment	6	8.900
Large home appliances	6	9.950
Television sets and monitors	5	4.800
IT and audio-video equipment	5	2.850
Small home appliances and others minor equipment	3	3.500
Total	**25**	**30.000**

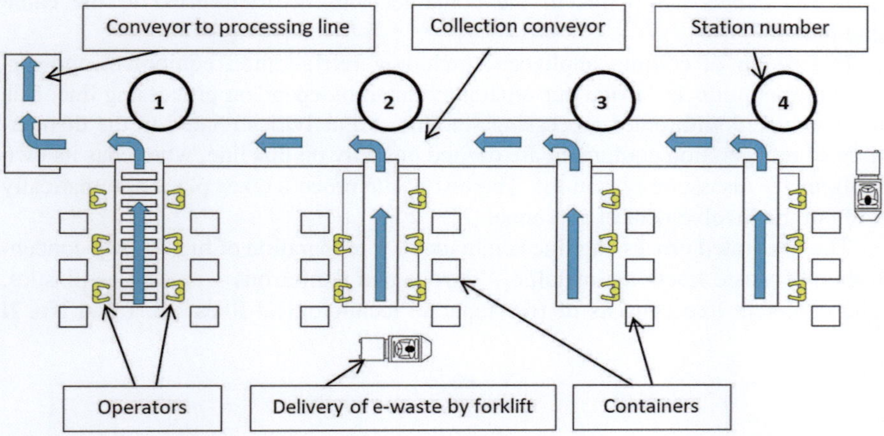

Fig. 2.4 View of the manual disassembly sector for different groups of equipment (1, large appliances; 2, information technologies (IT) equipment; 3, small home appliances; 4, TV sets and monitors)

In the first station, employees disassemble large home appliances except cooling and air-conditioning appliances (Fig. 2.4, 1). Components containing hazardous substances, electric motors, or counterweights (washing machines) will be placed in special containers. Power cables will be cut off. Disassembled components will be placed on belt conveyors. Then they will be transported to an inclined conveyor that supplies the material to the automated, integrated processing system No. I or No. II.

The second station (Fig. 2.4, 2) is for information technologies equipment disassembly – personal computers, servers, laptops, other peripherals, and audiovisual equipment, with the exception of televisions and monitors. It is important to preselect individual components like processors, memory cards, and mainboards containing precious metals and other printed circuit boards, covers, or power supply cables. High-grade classified components (including significant contents of gold, silver, or palladium) will be sold to third parties. Other disassembled components will be directed to integrated processing system No. I or No. II. The cables will be stored in separate containers for further processing at another machine in the plant.

The third station (Fig. 2.4, 3) disassembles the group of small-sized household appliances, electrical and electronic tools, as well as other small equipment. The operating time for disassembly will be reduced to a minimum due to a few basic operations such as cutting off the power cables and possible removal of elements containing dangerous substances (batteries, electrolyte condensers, etc.). Then, the waste equipment will be directed onto conveyor feeding the integrated processing system.

The preliminary disassembly of television sets and monitors is planned for the fourth station (Fig. 2.4). Flat screens after disassembly will be stored in separate containers for shipment to third-party plant. Other components, such as casing, printed circuit boards, etc., can be directed to the integrated system's processing line. The cables and wires in the container will be transported to the cable granulator.

Processing of cooling appliances including refrigeration equipment and air-conditioning units is carried out separately on an independent processing line. It is not associated with other processing stations. All activities related to the dismantling of refrigeration equipment are carried out only on this line, which has its own preliminary disassembly stations. The rest of the process takes place automatically without the involvement of personnel (Fig. 2.5).

The integrated processing line is intended for preparation of high-purity concentration of waste fractions: metallic – ferrous and nonferrous – as well as plastics. The complete line consists of two separate technological lines No. I and No. II

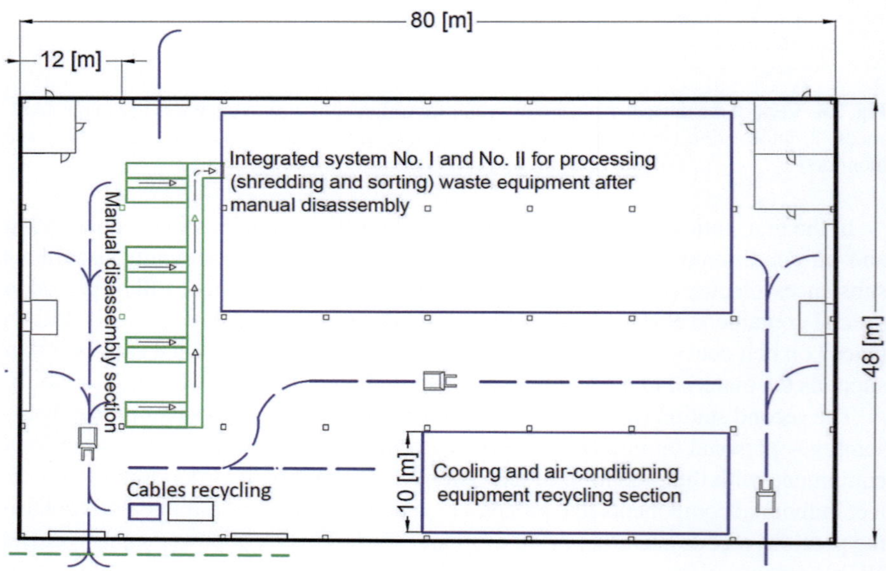

Fig. 2.5 The layout of the E-waste recycling plant with four main sections (manual disassembly station, integrated recycling system, cooling appliances recycling, cables granulator)

(Hamos 2014). Processing capacity is 2–4 [tonnes/h] for large home appliances and heavy equipment and 1–2 [tonnes/h] for small or medium equipment. Line No. I is equipped with (Hamos 2014):

- Belt conveyors and vibratory feeders
- Chain crusher
- Dust extraction systems
- Pneumatic sieves
- Separators: magnetic, eddy current, and electrostatic

The waste equipment and large assemblies after manual preliminary disassembly are transferred to the chain crusher. After crushing, the material transfer is with a vibrating feeder to pneumatic sieves and screens. The processing system is enclosed in a casing, so that dust formed during the process can be led outside by a pneumatic de-dusting system. The fraction of the processed material with larger dimensions not separated by pneumatic sieves goes to the manual sorting station, from which the components that have not been shredded can be shipped to other plants.

Line No. II is designed to process smaller electronic devices and equipment but also to crush residues from the No. I system and other sources. A hammer mill is used for grinding, in order to facilitate the separation of the metal fraction from other materials at a later stage. Then the waste is screened and dust is removed. The basic devices in line No. II include:

- Belt conveyors and vibratory feeders
- Hammer mill
- Screening separators
- Electrostatic, magnetic, and eddy current separators
- Gravity separators

The processing steps in line No. II are similar to No. I line, in addition to the manual sorting station for the separation of unprocessed components. The technical parameters of the entire system are summarized in Table 2.6. The basic version of the system stores material fractions after separation in elastic containers (big bags). It is possible to install conveyors for the transportation of processed material to containers placed outside the hall.

The equipment containing refrigerants require a different approach than other large home appliances. The refrigerant must be removed from each piece of waste – especially from the compressing unit. Disassembly takes place according to the following steps: after removing the shelves, drawers, glass, and other easily removable elements followed by refrigerant removal with the refrigerant suction system. Then the aggregate, evaporator, and condenser are removed. The first grinding is performed on a double shaft shredder, the second on a vertical chain crusher. The polyurethane foam is drawn off pneumatically using a cyclone separator. Separation of ferrous metals takes place on a magnetic separator and then on eddy current separator separates nonferrous metals from plastics. For technical data for the cooling appliances, recycling line is summarized in Table 2.7 (Vary 2018).

Table 2.6 Technical parameters of the integrated processing line

Productivity	No. I: 2–4 [t/h]; No. II: 1–2[t/h]
Power consumption	No. I: 350 kW
	No. II: 300 kW
The total price of the processing line	2.000.000 [EUR]

Table 2.7 Technical parameters of the cooling appliances processing line

Productivity	25–40 pcs/h
Power consumption	310 kW
The purity of the ferrous and nonferrous metals	>95%
The purity of plastics fraction	>90%
The total price of the processing line	500.000 [EUR]

Table 2.8 Main parameters of the cable granulator

Productivity	180–200 kg/h
Power consumption	22 kW
The purity of the output material	98–99,9%
Price of cable granulator	40.000 [EUR]

The cables after cutting them off from various types of equipment can be processed by cable granulators. The material, when inserted in the machine hopper, goes inside. After initial grinding, it is transferred by means of a suction system to a turbine mill, which had a material for fine granulator. Separation is carried out by gravity. The resulting dust is sucked into the filtering tube. The basic technical parameters of the cable granulator are shown in Table 2.8 (IRS 2018).

2.5 Case Study: Configuration of Small and Large Appliances Processing Line

Large home appliances disassembly starts with manual operations discussed in the previous section. Small home appliances potential scrap material value per piece is low. The cost of labor is too high in developed countries for conducting manual disassembly. Therefore, opening the casing of any small appliance should be processed mechanically for subsequent removal of any components containing hazardous substances. Figure 2.6 shows a different approach in a configuration of processing large home appliances and small home appliances. Cooling appliances and television sets are excluded from processing in this configuration. The processing line consists of two subordinate lines, A and B, which are connected at point C shown in Fig. 2.6. Large home appliances excluding cooling equipment manual disassembly are conducted at the station – 1 supplied by forklifts. Then the casings

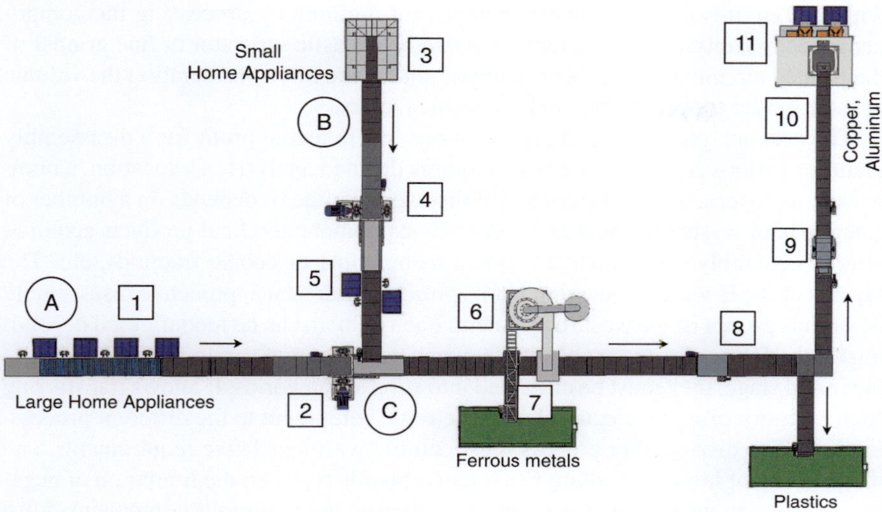

Fig. 2.6 The layout of reconfigurable processing lines for small and large home appliances (SHA and LHA). Description of the numbers is included in the text

are transferred to the shredder – 2. Line B is for processing small home appliances. It can be loaded by a bucket loader or a handling machine equipped with hydraulic grip inside a hopper – 3. A shredder – 4 is equipped with large wide teeth. The intention is to shred small equipment in a larger piece of materials to allow manual separation of components containing hazardous substances or homogeneous material parts at the station – 5. Both lines connect at point C. A cyclone separator – 6 is for removing dust or any light objects from further treatment. Further process includes magnetic separation – 7, eddy current separation – 8, and transfer of nonferrous metals and alloys to a hammer mill – 9. The final stage of separation is conducted on screening separator – 10 (size of grains) and gravity separators – 11 (light and heavy metal fraction). The line productivity can be adjusted by changing the number of manual disassembly stations accompanied by the increased capacity of individual processing machines.

2.6 Conclusions

A big variety of groups and types of waste equipment require preselection of the equipment similar in disassembling sequence. The findings of this study show that it is necessary to reduce a number and types of specialized shredding and sorting machines applied to disassembly lines. A modularity of a recycling plant by selection of autonomous units linked together to form a processing line would benefit in lower costs and the size of a plant. The concept of a reconfigurable recycling system of E-waste presented in this research assumes modularity of the disassembly line with a possibility of using other machines for preliminary processing the components and assemblies. As an example, it can be a plastic separator or fine grinder of the printed circuit boards. It is very important to take into consideration the volume of the E-waste for processing and economic factors.

The cost analyses indicated big variations in a potential profit for a disassembly plant for various countries. A design requires detailed analysis of a location, a number of employees, and configuration of disassembly line. It depends on a number of categories of waste electrical and electronic equipment and final products acquired after disassembly – components, mixed metals, fine or coarse fractions, etc. The layouts of the E-waste processing plants present a practical approach to disassembly of various groups of the waste equipment. Each of them can be reconfigured depending on the focus of disassembly strategy and estimation of economic indicators. At the initial stage, they must be classified into several categories. It allows transferring each category of waste electrical and electronic equipment to the different processing line. The disassembly process must comply with legislative requirements, and the majority of laws concerning E-waste worldwide focus on the limitation of negative impacts on the natural environment. Therefore the hazardous components must be removed from the stream of the material fractions in disassembly plant.

In developed countries, disassembly should focus on automation and optimization of the process, taking into consideration high labor costs. Such proposals were discussed, i.e., by automation of the process proposed by Gerbers and Alvarez-de-los-Mozos and Renteria (Gerbers et al. 2016) (Alvarez-de-los-Mozos and Renteria 2017). Studies with the concepts and case studies of automated or robotic disassembly show examples for liquid crystal displays disassembly (Elo and Sundin 2014; Vanegas Pena et al. 2016), the printed circuit boards (Park et al. 2015), and laptops (DiFilippo and Jouaneh 2017). Possible solutions for modular recycling system were discussed by Barwood et al. with a case study of a robotic disassembly cell (Barwood et al. 2015).

At the same time, it is a producer's responsibility to provide sufficient information about materials and substances used in the products to facilitate the sequence of disassembly when a product reaches the end of life. The information should include material inventory. Precious or rare earth metals and also hazardous substances should be identified (Nowakowski 2018). Some authors propose additional data to be used in new products. They propose disassembly steps for each electrical and electronic product (Huang et al. 2012; Wang et al. 2013; Song et al. 2014). Some proposals include an attachment of a special information matrix to facilitate and improve materials recycling (Luttropp and Johansson 2010).

In the developing countries, labor costs are lower in developed countries. Therefore, the main goal of recycling plants should focus on minimizing the negative impact on the environment by removing all hazardous substances from the waste equipment. We should take into consideration capabilities of recycling individual elements, especially metals, in a country where disassembly is conducted. In such case, all components containing specified above materials should be safely shipped to other plants for recycling and refining individual element or for neutralizing hazardous substance.

Future work should be focused on the development of identification methods and automated disassembling units. It is necessary to work out robotic sections where the operations of hazardous substances removal could be conducted without manual work dangerous for employees.

References

ACRR (2009) The management of WEEE. A guide for local and regional authorities. Association of cities and regions for recycling

Alvarez-de-los-Mozos E, Renteria A (2017) Collaborative robots in e-waste management. Procedia Manuf 11:55–62. https://doi.org/10.1016/j.promfg.2017.07.133

Bakar MSA, Rahimifard S (2008) Ecological and economical assessment of end-of-life waste recycling in the electrical and electronic recovery sector. Int J Sustain Eng 1:261–277. https://doi.org/10.1080/19397030802576825

Barba-Gutiérrez Y, Adenso-Díaz B, Hopp M (2008) An analysis of some environmental consequences of European electrical and electronic waste regulation. Resour Conserv Recycl 52:481–495. https://doi.org/10.1016/j.resconrec.2007.06.002

Barwood M, Li J, Pringle T, Rahimifard S (2015) Utilisation of reconfigurable recycling systems for improved material recovery from E-waste. Procedia CIRP 29:746–751. https://doi.org/10.1016/j.procir.2015.02.071

BMAS (2018) Bundes Ministerium Fur Arbeit und Sozial Germany. Mindestlohn-Rechner (Minimum wage calculator). http://www.bmas.de/DE/Themen/Arbeitsrecht/Mindestlohn/Rechner/mindestlohnrechner.htm. Accessed 21 Feb 2018

Bogaert S, Van Acoleyen M, Van Tomme I, et al (2008) Study on RoHS and WEEE Directives. ARCADIS & RPA for the DG ENTR, European Commission

Capraz O, Polat O, Gungor A (2015) Planning of waste electrical and electronic equipment (WEEE) recycling facilities: MILP modelling and case study investigation. Flex Serv Manuf J 27:479–508. https://doi.org/10.1007/s10696-015-9217-3

Capraz O, Polat O, Gungor A (2017) Performance evaluation of waste electrical and electronic equipment disassembly layout configurations using simulation. Front Environ Sci Eng 11(5)

CENELEC (2013) European standard EN 50625–1 – collection, logistics & treatment requirements for WEEE – part 1: general treatment requirements

Chung S-W, Murakami-Suzuki R (2008) A comparative study of e-waste recycling systems in Japan, South Korea and Taiwan from the EPR perspective: implications for developing countries. http://www.ide.go.jp/English/Publish/Download/Spot/pdf/30/007.pdf. Accessed 19 Oct 2011

Chung C, Peng Q (2005) An integrated approach to selective-disassembly sequence planning. Robot Comput Integr Manuf 21:475–485. https://doi.org/10.1016/j.rcim.2004.11.008

Chung S-S, Zhang C (2011) An evaluation of legislative measures on electrical and electronic waste in the People's Republic of China. Waste Manag 31:2638–2646. https://doi.org/10.1016/j.wasman.2011.07.010

Cooper T (2004) Inadequate life? Evidence of consumer attitudes to product obsolescence. J Consum Policy 27:421–449. https://doi.org/10.1007/s10603-004-2284-6

Cucchiella F, D'Adamo I, Lenny Koh SC, Rosa P (2016) A profitability assessment of European recycling processes treating printed circuit boards from waste electrical and electronic equipments. Renew Sust Energ Rev 64:749–760. https://doi.org/10.1016/j.rser.2016.06.057

DiFilippo NM, Jouaneh MK (2017) A system combining force and vision sensing for automated screw removal on laptops. IEEE Trans Autom Sci Eng 15:887–895

Duta L, Filip G, Henrioud JM, Popescu C (2008) Disassembly line scheduling with genetic algorithms. Int J Comp, Commun Control 3:230–240

Dwivedy M, Mittal RK (2012) An investigation into e-waste flows in India. J Clean Prod 37:229–242. https://doi.org/10.1016/j.jclepro.2012.07.017

Elo K, Sundin E (2014) Automatic dismantling challenges in the structural design of LCD TVs. Procedia CIRP 15:251–256

ElSayed A, Kongar E, Gupta SM, Sobh T (2012) A robotic-driven disassembly sequence generator for end-of-life electronic products. J Intell Robot Syst 68:43–52. https://doi.org/10.1007/s10846-012-9667-8

European Commission (2012) Directive 2012/19/EU of the European Parliament and of the council of 4 July 2012 on waste electrical and electronic equipment (WEEE)text with EEA relevance – LexUriServ.do

European Union (2003) Directive 2002/96/EC Of The European Parliament and of The Council of 27 January 2003 on waste electrical and electronic equipment (WEEE). http://eur-lex.europa.eu/LexUriServ/LexUriServ.do?uri=OJ:L:2003:037:0024:0038:en:PDF. Accessed 28 Oct 2011

Eurostat (2018) Eurostat electricity price statistics – statistics explained. http://ec.europa.eu/eurostat/statistics-explained/index.php/Electricity_price_statistics. Accessed 28 Mar 2018

Gerbers R, Mücke M, Dietrich F, Dröder K (2016) Simplifying robot tools by taking advantage of sensor integration in human collaboration robots. Procedia CIRP 44:287–292. https://doi.org/10.1016/j.procir.2016.02.135

GRN (2018) Global recycling network – polystyrene waste exchange listings http://www.grn.com/a/view/1006.html. Accessed 22 Mar 2018

GUS (2018) Główny Urząd Statystyczny. Statistics Poland, Warsaw. Wiadomości Statystyczne 3 (682). https://stat.gov.pl/files/gfx/portalinformacyjny/pl/defaultstronaopisowa/5909/29/1/wiadomosci_statystyczne_03_2018.pdf. Accessed 21 Feb 2018

Hall WJ, Williams PT (2007) Separation and recovery of materials from scrap printed circuit boards. Resour Conserv Recycl 51:691–709

Hamos (2014) WEEE scrap recycling plant – Hamos ERP – ERP system. www.hamos.com/upload/pdf/brochures/product_range/hamos_Product_Range.pdf. Accessed 21 Feb 2014

Hischier R, Wager P, Gauglhofer J (2005) Does WEEE recycling make sense from an environmental perspective?:: the environmental impacts of the Swiss take-back and recycling systems for waste electrical and electronic equipment (WEEE). Environ Impact Assess Rev 25:525–539. https://doi.org/10.1016/j.eiar.2005.04.003

Horner R, Gertsakis J (2006) Literature review on the environmental and health impacts of waste electrical and electronic equipment. Report prepared for the Ministry for the Environment, Government of New Zealand

Hsin-Hao (Tom) Huang Y, Wang MH, Johnson MR (2000) Disassembly sequence generation using a neural network approach. J Manuf Syst 19:73–82

Huang C-C, Liang W-Y, Chuang H-F, Chang Z-Y (2012) A novel approach to product modularity and product disassembly with the consideration of 3R-abilities. Comput Ind Eng 62:96–107. https://doi.org/10.1016/j.cie.2011.08.021

Huisman J, Magalini F (2007) 2008 Review of directive 2002/96 on Waste Electrical and Electronic Equipment (WEEE)

Ikhlayel M (2017) Environmental impacts and benefits of state-of-the-art technologies for E-waste management. Waste Manag 68:458. https://doi.org/10.1016/j.wasman.2017.06.038

INSEE (2018) Earnings – since 1991, the monthly series. Statistical data and metadata databases – Romania http://www.insse.ro/cms/en/content/earnings-1991-monthly-series. Accessed 21 Feb 2018

IRS (2018) Macchine Compatte Trattamento Cavi – Kombi. http://www.irsitalia.net/media/1044/kombiline_-catalogo.pdf. Accessed 6 June 2014

Kahhat R, Kim J, Xu M et al (2008) Exploring e-waste management systems in the United States. Resour Conserv Recycl 52:955–964. https://doi.org/10.1016/j.resconrec.2008.03.002

Kang HY, Schoenung JM (2005) Electronic waste recycling: a review of US infrastructure and technology options. Resour Conserv Recycl 45:368–400

Kang H-Y, Schoenung JM (2006) Economic analysis of electronic waste recycling: modeling the cost and revenue of a materials recovery facility in California. Environ Sci Technol 40:1672–1680

Keri C (2012) 15 – recycling cooling and freezing appliances. In: Goodship V, Stevels A (eds) Waste electrical and electronic equipment (WEEE) handbook. Woodhead Publishing, London, pp 339–351

Kuo TC (2010) The construction of a collaborative-design platform to support waste electrical and electronic equipment recycling. Robot Comput Integr Manuf 26:100–108

Lambert AJD (2002) Determining optimum disassembly sequences in electronic equipment. Comput Ind Eng 43:553–575

Lambert AJD (2003) Disassembly sequencing: a survey. Int J Prod Res 41:3721–3759. https://doi.org/10.1080/0020754031000120078

Lee SG, Lye SW, Khoo MK (2001) A multi-objective methodology for evaluating product end-of-life options and disassembly. Int J Adv Manuf Technol 18:148–156

Lee J, Song HT, Yoo JM (2007) Present status of the recycling of waste electrical and electronic equipment in Korea. Resour Conserv Recycl 50:380–397. https://doi.org/10.1016/j.resconrec.2007.01.010

Li JR, Khoo LP, Tor SB (2006) Generation of possible multiple components disassembly sequence for maintenance using a disassembly constraint graph. Int J Prod Econ 102:51–65

Li J, Lopez NBN, Liu L et al (2013) Regional or global WEEE recycling. Where to go? Waste Manag 33:923–934. https://doi.org/10.1016/j.wasman.2012.11.011

Li J, Barwood M, Rahimifard S (2017) Robotic disassembly for increased recovery of strategically important materials from electrical vehicles. Robot Comput Integr Manuf 50:203–212

Liu X, Tanaka M, Matsui Y (2009) Economic evaluation of optional recycling processes for waste electronic home appliances. J Clean Prod 17:53–60

LME (2018) London metal exchange – metals – trading summary. London. https://www.lme.com/en-GB/Metals. Accessed 21 Mar 2018

Luttropp C, Johansson J (2010) Improved recycling with life cycle information tagged to the product. J Clean Prod 18:346–354. https://doi.org/10.1016/j.jclepro.2009.10.023

Nowakowski P (2015) Logistyka recyklingu zużytego sprzętu elektrycznego i elektronicznego. Od projektowania po przetwarzanie – Monografia. Wydaw. Politechniki Śląskiej

Nowakowski P (2016) The influence of residents' behaviour on waste electrical and electronic equipment collection effectiveness. Waste Manag Res 34:1126–1135

Nowakowski P (2017) Identyfikacja czynników wpływających na efektywność łańcucha logistyki zwrotnej zużytego sprzętu elektrycznego i elektronicznego. (in polish – identification of the factors having impact of the efficiency of the reverse supply chain of WEEE). In: Knosala R (ed) Innowacje w zarządzaniu i inżynierii produkcji. T. 2. Oficyna Wydaw. Polskiego Towarzystwa Zarządzania Produkcją, Opole, pp 65–78

Nowakowski P (2018) A novel, cost efficient identification method for disassembly planning of waste electrical and electronic equipment. J Clean Prod 172:2695–2707. https://doi.org/10.1016/j.jclepro.2017.11.142

Oguchi M, Murakami S, Sakanakura H et al (2011) A preliminary categorization of end-of-life electrical and electronic equipment as secondary metal resources. Waste Manag 31:2150–2160. https://doi.org/10.1016/j.wasman.2011.05.009

Park S, Kim S, Han Y, Park J (2015) Apparatus for electronic component disassembly from printed circuit board assembly in e-wastes. Int J Miner Process 144:11–15. https://doi.org/10.1016/j.minpro.2015.09.013

Rosa P, Terzi S (2016) Comparison of current practices for a combined management of printed circuit boards from different waste streams. J Clean Prod 137:300–312. https://doi.org/10.1016/j.jclepro.2016.07.089

Santochi M, Dini G, Failli F (2002) Disassembly for recycling, maintenance and remanufacturing: state of the art and perspectives. AMST '02 Adv Manuf Syst Technol 1:73–89

Sarath P, Bonda S, Mohanty S, Nayak SK (2015) Mobile phone waste management and recycling: views and trends. Waste Manag 46:536–545. https://doi.org/10.1016/j.wasman.2015.09.013

Song X, Zhou W, Pan X, Feng K (2014) Disassembly sequence planning for electro-mechanical products under a partial destructive mode. Assem Autom 34:106–114. https://doi.org/10.1108/AA-01-2013-006

SWI (2018) SWI swissinfo.ch Misicka S, Reusser K. What do people earn in Switzerland? https://www.swissinfo.ch/eng/business/salaries_what-do-people-earn-in-switzerland/43035438. Accessed 21 Feb 2018

SWISSGRID (2018) Tariffs. https://www.swissgrid.ch/dam/swissgrid/experts/grid_usage/Tabelle_Tarife_en.pdf. Accessed 28 Mar 2018

Tanskanen P (2012) Electronics waste: recycling of mobile phones. In: Post-consumer waste recycling and optimal production, pp 129–150

Terazono A, Oguchi M, Iino S, Mogi S (2015) Battery collection in municipal waste management in Japan: challenges for hazardous substance control and safety. Waste Manag 39:246–257

Vanegas Pena P, Peeters J, Cattrysse D, et al (2016) Study for a method to assess the ease of disassembly of electrical and electronic equipment. Method development and application to a flat panel display case study Publications Office of the European Union. https://doi.org/10.2788/130925

Vary (2018) Refrigerator harmless treatment & resource recycling system – Varygroup. http://www.varygroup.com/product/9. Accessed 6 Jan 2018

Wang H, Xiang D, Rong Y, Zhang L (2013) Intelligent disassembly planning: a review on its fundamental methodology. Assem Autom 33:78–85. https://doi.org/10.1108/01445151311294801

Wath SB, Dutt PS, Chakrabarti T (2011) E-waste scenario in India, its management and implications. Environ Monit Assess 172:249–262. https://doi.org/10.1007/s10661-010-1331-9

Williams JAS (2006) A review of electronics demanufacturing processes. Resour Conserv Recycl 47:195–208. https://doi.org/10.1016/j.resconrec.2005.11.003

Chapter 3
An Economic Assessment of Present and Future Electronic-Waste Streams: Japan's Experience

Hitoshi Hayami and Masao Nakamura

Contents

Abstract In this chapter, we discuss some of the most important factors, including legal, statistical, economic, and organizational factors, that affect the recycling of waste electrical and electronic equipment or more broadly the recycling of general Electronic-waste in Japan and other countries. In doing so, we emphasize the policy importance of incorporating manufacturing supply chains in the design of environmental management of production systems.

The research reported here is in part supported by the Keio Gijuku Academic Development Funds and the Social Sciences and Humanities Research Council of Canada.

H. Hayami
Faculty of Business and Commerce, Keio University, Tokyo, Japan

M. Nakamura (✉)
Sauder School of Business, University of British Columbia, Vancouver, BC, Canada

Graduate School of Business and Commerce, Keio University, Tokyo, Japan
e-mail: masao.nakamura@sauder.ubc.ca

A. Khan et al. (eds.), *E-waste Recycling and Management*,
Environmental Chemistry for a Sustainable World 33,
https://doi.org/10.1007/978-3-030-14184-4_3

We also point out that the rates of collecting and recycling waste electrical and electronic equipment are relatively low in Japan as well as in the European Union countries. This chapter puts forward some recommendations that need to be taken into account in the public policy debate in order that the current low rates are to be improved.

3.1 Introduction

Much attention has been paid to the generation of Electronic-waste (typically termed as "E-waste") in many countries in recent years. Electronic-waste, for example, represents waste electrical and electronic equipment (typically termed as "WEEE"), among other product categories, and is measured in various ways.

The amounts of electronic-waste generated globally have been increasing over time. For example, Fig. 3.1 shows that per capita generation of electronic-waste in Asian countries has been increasing in recent years. In fact, in a growing economy, the total generation of electronic-waste is likely to continue increasing (Kusch and Hills 2017).

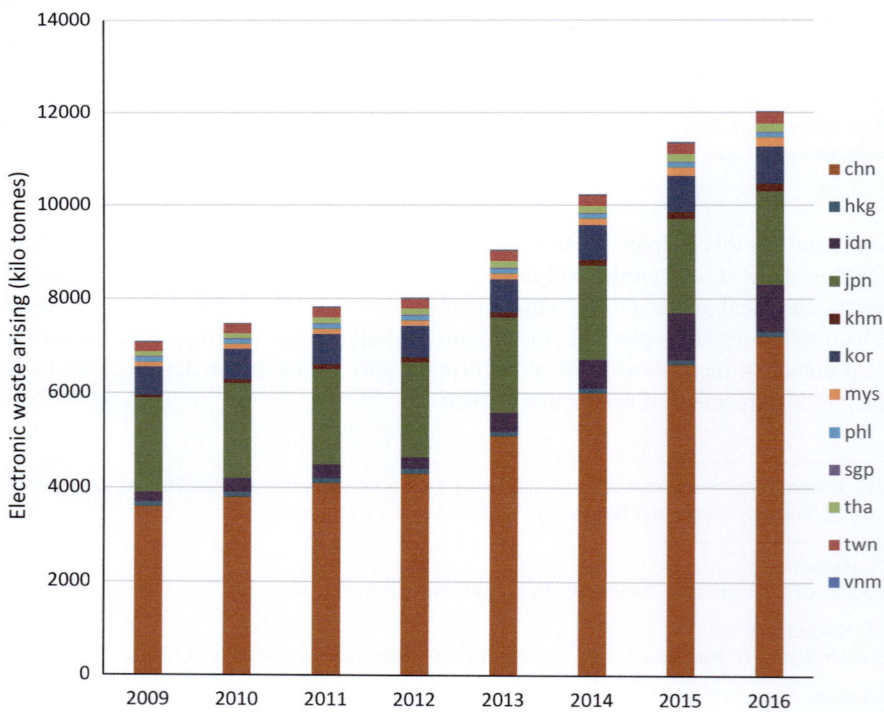

Fig. 3.1 The growth of Electronic-waste in East and Southeast Asia. (Source: United Nations University: Baldé et al. (2015))

The United Nations reports that a record level of waste electrical and electronic equipment, amounting to 41.8 million tons worldwide, was thrown away in 2014, with less than one-sixth of it being properly recycled (Baldé et al. 2015, page 24). It was the largest amount ever of Electronic-waste that was discarded, and there is little sign of a slowdown. Even countries that have recycling and recovery programs including Japan discard large amounts of waste electrical and electronic equipment.

The largest amount of Electronic-waste was generated in the United States and China, which together accounted for 32% of the total. The third most wasteful country by volume was Japan, which discarded a grand total of 2.2 million tons in 2013 (Japan Times, 2015, May 9).

Even though Japan's per capita waste, 17.3 kg per inhabitant, was lower than some less densely populated countries, other countries, such as those in Africa, had much lower amounts of Electronic-waste. Africa's average was 1.7 kg per person, one-tenth the amount of the waste generated by the average Japanese.

This kind of refuse is dangerous and often highly toxic. The refrigerators, washing machines, and microwave ovens routinely discarded contain large amounts of lead glass, batteries, mercury, cadmium, chromium, and other ozone-depleting chlorofluorocarbons (often termed as "CFCs"). The 7% of Electronic-waste last year made up of mobile phones, calculators, personal computers, printers, and small information technology equipment also contained poisonous components.

Electronic-waste last year also contained valuable resources worth $52 billion, only a quarter of which was recovered. Worldwide, an estimated 16.5 million tons of iron, 1.9 million tons of copper, 300 tons of gold (equal to 11% of the world's total gold production in 2013), as well as silver, aluminum, and palladium plastic were simply thrown out. With better recovery systems, those resources wouldn't end up in dumps, increasingly located in poorer countries, but would be recycled.

Japan was one of the first countries to impose recycling of Electronic-waste, and the Japanese system is thought to be better than in many countries. However, Japan still only treats around 24–30% of its Electronic-waste, the report estimated. The Japanese government reported that 556,000 tons of Electronic-waste was collected and treated in Japan in 2013, but that still only accounts for one-quarter of the total.

The convenience people have sought in the kitchen, laundry, and bathroom, and for daily communication, has become the world's noxious waste. With rising sales and shorter life cycles for products, the Electronic-waste problem is not likely to improve anytime soon.

Individuals should make sure that their disposal of even small gadgets is handled correctly. Governments around the world, including Japan, need to impose stricter rules, establish better disposal and recycling systems, and increase oversight.

One of the main reasons that per capita Electronic-waste generally grows over time is that per capita Electronic-waste increases with per capita gross domestic product (typically termed as "GDP"). Kusch and Hills (2017) present evidence that there is a positive correlation between these two quantities observed cross-sectionally for many nations in the Pan-European region.[1] Furthermore, this correlational relationship seems to hold regardless of the stages of economic development of specific countries in their sample.

Current Issues

As we noted above, the generation of Electronic-waste is likely to continue to grow over time globally. The often included items in waste electrical and electronic equipment are air conditioners, refrigerators, washing machines, television sets, other appliances, and cell phones. Many of these items are bulky and difficult to dispose of. In addition, they might produce toxic substances on the grounds if abandoned. From the life cycle perspectives,[2] production of these products requires large amounts of metal, energy, and other resources and generates significant amounts of greenhouse gases (CO_2, CH_4, N_2O, and CFCs). For these and other reasons for recycling of waste electrical and electronic equipment was identified as an important issue of environmental management.

Recycling of Waste Electrical and Electronic Equipment

While many countries recognize that promoting recycling of waste electrical and electronic equipment is an important policy issue, there are still a number of difficulties to implement policies that promote such recycling. For example, since the items included in waste electrical and electronic equipment are generally consumer goods, such policies must be compatible with consumers' incentives. Similarly, we would like to include producers and/or retailers of these products in the recycling process since retailers, for example, will have first-hand information on the customers who purchase these products. Delegating some responsibility of recycling to producers and retailers is called extended producer responsibility (EPR) and is often included in waste electrical and electronic equipment recycling laws in many countries.

Economics of Recycling Waste Electrical and Electronic Equipment

It is essential to pay attention to the economic principles underlying policy matters on environmental management such as collecting, recycling, and processing of Electronic-waste. Since the products which generate Electronic-waste after they are consumed are produced using metals and also precious metals for some products like cell phones, recovering some of these metals and precious metals from recycled waste electrical and electronic equipment items is likely to give some economic benefits.[3] Also, collecting discarded waste electrical and electronic equipment from consumers' homes and transporting them to the specified deposit is not free. For

---

[1] Kusch and Hills (2017) show in particular that an increase of 1000 international $ GDP purchasing power parity (PPP) means an additional 0.5 kg waste electrical and electronic equipment is generated.

[2] We regard environmental management analyses and approaches based on life cycle and supply chain perspectives interchangeably (see, e.g., Hayami et al. (2015) and Hayami and Nakamura (2007)).

[3] This is discussed in Sect. 3.3.

Table 3.1 Cost and benefit analysis of waste electrical and electronic equipment recycling: consideration of life cycle (supply chain) stages

Life cycle stages	Cost/benefit implications for recycling process Examples
(1) Design and production stage	Amounts of metals and precious metals used (greenhouse gas emissions/transportation costs of the collection of recycled items)
(2) Sales/marketing stage	Contractual arrangements for expected recycling activities (how the items are to be collected when product life is over)
(3) Collection of used products	Logistics for collection/locations for recycled products (fuel costs of transportation)

Source: By the authors

example, in Japan, these items typically weigh as follows: television set (28 kg/unit), air conditioner (43 kg/unit), refrigerator (58 kg/unit), and washing machine (32 kg/unit). Collecting and transporting of these used appliances is certainly costly, and their costs need to be compared with the economic benefits obtained from recycling.

Stages of Recycling Processes of Waste Electrical and Electronic Equipment
The above discussion suggests that the recycling processes must be studied analytically as follows (Table 3.1), taking into account the effects of recycling-related costs as well as direct and indirect costs and benefits from the recycling.

For example, designing products using smaller amounts of metals and simple designs may allow recycling costs to be reduced, as well as reductions in fuel costs and the associated emissions of greenhouse gases (typically termed as "GHG"). These cost reductions may outweigh the benefits of recovering some marketable metals in the recycling process. Contractual arrangements between producers and customers (consumers) might matter in order to increase the recycling rates of waste electrical and electronic equipment. Similarly, an optimal spatial distribution of the locations for the collecting depot points of recycled products may also facilitate reductions in the cost of transportation.

3.2 Electronic-Waste Management in Japan

Japan was among the first countries which began Electronic-waste recycling. Because of the rapid technological changes that took place in the areas of production and consumption of waste electrical and electronic equipment items recently in Japan, it is of considerable academic and practical interest to study the recycling and other activities related to waste electrical and electronic equipment in Japan. In this section, we describe the basic legal institutions (laws) that oversee Electronic-waste recycling activities in Japan, and then we discuss policy issues related to them.

3.2.1 Legal Institutions Overseeing Electronic-Waste Management in Japan

Two basic laws that oversee environmental management policies in Japan were put forward in 1994 and 1998, respectively. We briefly discuss these laws below.

- 1994: *Law for the Promotion of Effective Utilization of Resources.* This law was subsequently enacted in May 2000 and was put into force in April 2001.

  This law aims at establishing a sound material-cycle economic system by:

  (i) Enhancing measures for recycling goods and resources by implementing the collection and recycling of used products by business entities
  (ii) Reducing waste generation by promoting resource saving and ensuring longer life of products
  (iii) Newly implementing measures for reusing parts recovered from collected used products and at the same time as measures to address the reduction of industrial wastes by accelerating the reduction of by-products and recycle

This is an epoch-making law which requires to reduce, reuse, and recycle (typically termed as "3Rs") as part of measures; covers from upstream part, including product design; and measures against industrial wastes through downstream part such as collection and recycling of used products.[4]

- 1998, 2001: *Home Appliance Recycling Act* became a law in June 1998, but it became operational much later in 2001.

The primary objective of this *Home Appliance Recycling Act* is to operationalize its policy contents. It states that "This legislation shall have the objective of contributing to the maintenance of the living environment and the healthy development of the national economy, by taking steps to secure the environmentally sound disposal of waste and effective utilization of resources through the introduction of measures for proper and smooth collection, transportation, and recycling of specific household appliance waste by retail traders or manufacturers of specific household appliances, with the aim of achieving a reduction in the volume of general waste and sufficient utilization of recycled resources."

More specifically, for achieving this objective, this Act is designed to solve the following problems:

(i) Environmentally sound disposal of wastes (hazardous wastes) waste electrical and electronic equipment that is disposed of as bulky waste contains hazardous materials and pollutants. These include chlorofluorocarbons as both greenhouse gas and ozone-depleting substance, oil in motors and compressors, and heavy metals used in making printed circuit boards. Illegal dumping of such products poses even greater environmental risks. Thus, a system to manage waste electrical and electronic equipment in an environmentally sound manner was

[4] See Japanese Ministry of Justice (2017) and Japanese Ministry of Environemnt (2001).

expected to be built. In addition, since environmentally sound management of these wastes was often beyond the capacity of individual local governments, the manufacturers of these appliances were expected to participate in the process of managing these wastes.

(ii) Effective use of recyclable materials waste electrical and electronic equipment contains large amounts of iron, aluminum, copper, and glass. These can be an effective source of materials if they can be recovered efficiently.

The target areas of this act are the following four categories of home appliances:

1. Air conditioners
2. Television sets (cathode-ray tubes) and liquid crystal display types, excluding those designed to be incorporated into a building and that do not use primary batteries or storage batteries for their power source, as well as the plasma types
3. Electric refrigerators and freezers
4. Electric washing machines and clothes dryers

Also, flat-screen television sets (liquid crystal display and plasma types) and clothes dryers were added to the designated categories in April 2009.

Among other typical waste electrical and electronic equipment items, personal computers are managed under the previously discussed act called *Act on the Promotion of Effective Utilization of Resources (1994)*. Also, small electronic appliances such as mobile phones have been managed under a new law called *Small Electrical and Electronic Equipment Recycling Act* since 2013.

3.2.2 Overview of Electronic-Waste Recycling

Environmental management policies need to focus on, among other topics, (i) recycling of Electronic-waste, particularly its costs aspects, and its implications for (ii) reductions (if any) in the generation of greenhouse gases and (iii) reductions in the use of resources such as metal and precious metals. Earlier we discussed cost issues associated with the transportation of recycled Electronic-waste. How do such recycling costs compare with the tangible benefits of recycling (e.g., the commercial value of metals recycled, etc.)? Such cost and benefit trade-offs and analyses may ultimately determine the publicly justifiable degree of recycling activities.

From Table 3.2, we see that Japan's per capita Electronic-waste generated is considerably higher than that of Germany. Japan collects and recycles about a fourth of its per capita Electronic-waste generated.

In later sections, we will further look at the European Union's performance in Electronic-waste recycling in comparison with Japan's. Figure 3.2 shows the general upward trend in Japan's waste electrical and electronic equipment over time. It also shows a volatile pattern in the collection and recycling of total waste electrical and electronic equipment units. This suggests the importance of public policies that facilitate more robust performance in recycling activities.

Table 3.2 Generation and collection/recycling of Electronic-waste: Germany and Japan, 2013

	Germany	Japan
Electronic-waste generated (per capita), in kg	21.6	17.3
Total Electronic-waste generated, in kilotons	1769	2200
Total Electronic-waste collected and recycled, in kilotons	–	556
Population	81,589	127,061
National regulation of Electronic-waste recycling	Yes	Yes

Source: Compiled by the authors based on publicly available information. JEMAI (2017), and EU Eurostat (2017)

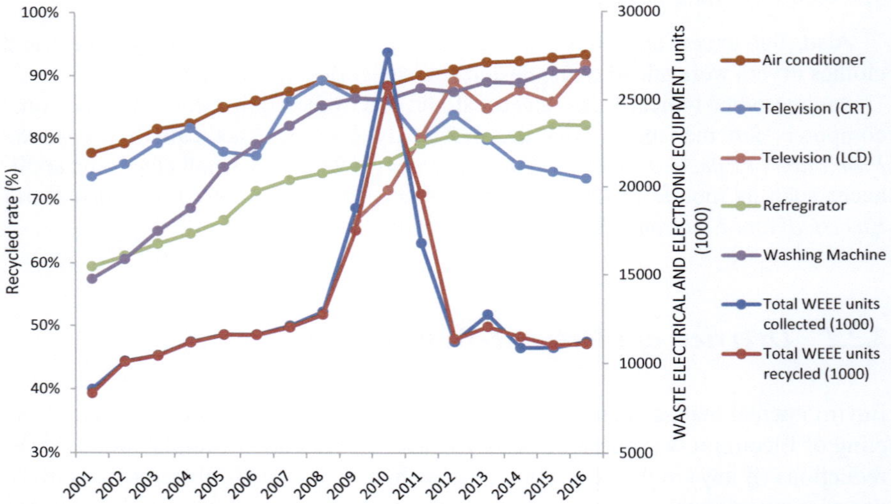

Fig. 3.2 Rates (%) of recycling and collection of Electronic-waste in Japan, 2001–2016. (Source: Compiled by the authors based on data from the Japanese Ministry of Environment Data Source http://www.env.go.jp/policy/keizai_portal/A_basic/a06.html and other sites)

3.2.3 Costs of Electronic-Waste Recycling

Given the large numbers and weights of units of these products in waste electrical and electronic equipment that need to be collected physically for recycling,[5] it is not difficult to see that the transportation costs play an important role among the deter-

[5] See Sect. 3.3 for further information regarding the physical border of waste electrical and electronic equipment items to be recycled.

minants of the recycling rates. (See Sect. 3.3 for the numbers of Electronic-waste items that need to be collected for recycling.)

The observed recycling costs of home appliances in Japan vary significantly depending on the types of appliance products to recycle, the companies which provide services to remove appliances out of homes and to transport them to the recycling depots, among other things, and also the availability of local recycling services provided by the local government offices. Table 3.3 presents a few examples of such

Table 3.3 Home appliances recycling costs: some examples, Japan, 2015

Product	Recycling fee (in yen)	Transportation cost (in yen)	Cost of removal from home (in yen)
Products covered by the 2001 Home Appliance Recycling Act			
Air conditioner	972[a]	500–3000	3000–20,000
TV sets (with screens below 15″)	1836	500–3000	3000–20,000
TV sets (with screens larger than 16″)	2916	500–3000	3000–20,000
Refrigerators (capacity below 170 liters)	3672	500–3000	3000–20,000
Refrigerators (capacity above 171 liters)	4644	500–3000	3000–20,000
Washing machine	2484	500–3000	3000–20,000
Personal computer	3000–4000[b]		
Other appliances	**Recycling fee (in yen, Tokyo 23 wards government offices)**	**Recycling fee (in yen, private recycling businesses)**	**Transportation and removal costs (in yen, private recycling businesses)**
Oil heater	700	1500 or more	Yes
Audio equipment	300	1000 or more	Yes
Gas burner	300	1000 or more	Yes
Lighting equipment	300	700 or more	Yes
Dishwasher/dryer	1000	1500 or more	Yes
Electric fan	300	500 or more	Yes
Microwave oven	300	800 or more	Yes
Video recorder	300	1000 or more	Yes
Printer	300–1000	1000 or more	Yes
Home sewing machine	700	2800 or more	Yes

Notes:

[a]Monetary figures presented in this table are in Japanese yen in 2016. The exchange rate between the Japanese yen and the US dollar for December 2016: \$1.00 = 116 Japanese yen

[b]Personal computers with publicly registered marks for recyclable products may be turned into their producers for recycling for free. Other public organizations may collect used personal computers for recycling free of charge. But the private sector businesses usually charge recycling fees

Source: Compiled by the authors based on public-use government information from Enechange (2016)

Table 3.4 Numbers of Electronic-waste items collected at designated sites across Japan, 2016

Product (Electronic-waste item)	Number of collected appliances		
	#Collected appliances (in 1000s of units)	Fraction of total (%)	Year-on-year change
Air conditioners	2567	22.9	+9.0%
CRT TVs	1184	10.6	−23.7%
LCD and plasma TVs	1279	11.4	+23.8%
Refrigerators and freezers	2829	25.3	+1.1%
Clothes washers and dryers	3339	29.8	+6.4%

Note: The observation period is between April 2016 and March 2017 (Japan's 2016 fiscal year)
Source: Compiled by the authors based on information available from METI (2017)

recycling costs observed in Japan for certain home and other appliances. Most of these appliances were sold by large national appliance producers.

As we see from Table 3.4, the recycling rates (or collection rates here) of the main Electronic-waste products for 2016 for Japan were all less than 30% and are the only small fractions of the total products sold to the consumers. While we can think of many possible reasons for this, the costs of recycling given in Table 3.3, which are relatively high compared to the prevailing prices of equivalent new products in the markets, might be in part responsible.

3.3 Life Cycle Policy Analysis Using Input-Output (I-O) Tables: Recycling of Mobile Phones and Personal Computers and Their Supply Chains in Japan

So far we have not discussed recycling of mobile phones and personal computers. These products were not included in the original Japanese laws on Electronic-waste recycling as we discussed in the previous sections. In this section, we calculate the indirect savings in the use of materials and other resources resulting from the recycling and processing of these electronic products. To do this we use the input-output tables for the Japanese economy (e.g., Hayami et al. (2015), Hayami and Nakamura (2007)).

Trends in the recycling of mobile phones and personal computers are presented in Figs. 3.3, 3.4, and 3.5.

Rates of the recycling of mobile phones for the phone bodies, batteries, and chargers remain relatively stable over the recent years at levels below 10 million units (Fig. 3.3). This is despite the significant decline in the shipments of new phone units. On the other hand, the recycling patterns for both desktop and notebook personal computers show that the general decline and fluctuations in the shipments of these personal computers are also reflected in the rates of recycling for these products (Figs. 3.4 and 3.5).

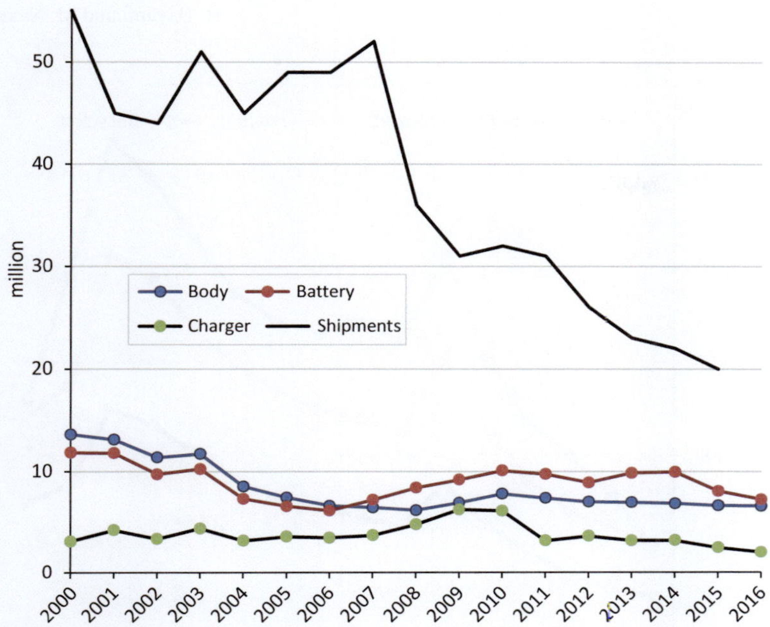

Fig. 3.3 Trend of recycled mobile phones by parts: body, battery, and charger compared to the shipments (in million). (Source: compiled by the authors using information for public use from Mobile Recycle Network (2018) and JEITA (2017).

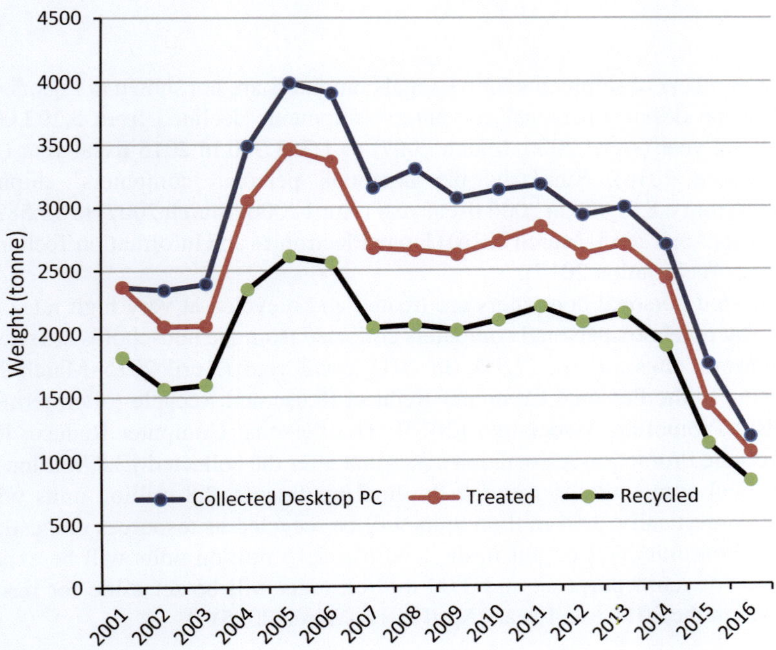

Fig. 3.4 Trend of the number of recycled desktop personal computers

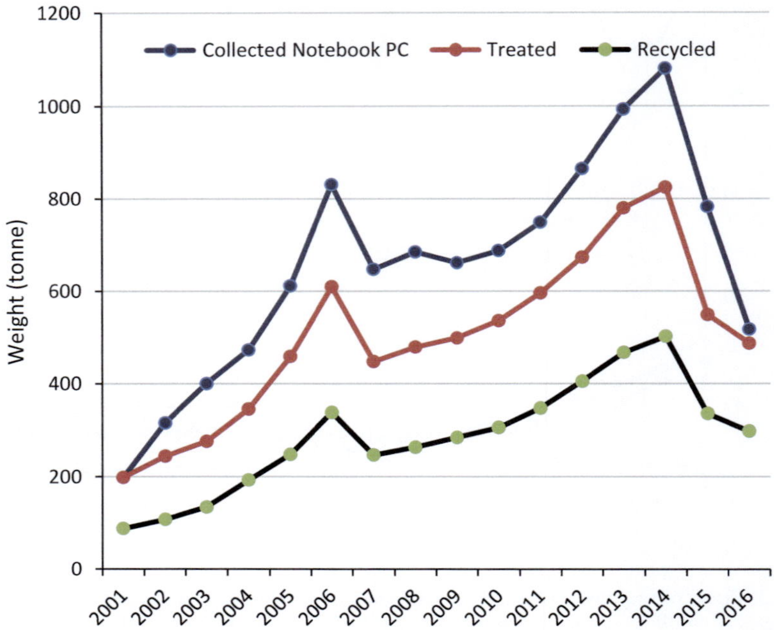

Fig. 3.5 Trend of the number of recycled notebook personal computers. (Source: compiled by the authors using information for public use from PC3R Promotion Association (2017)

The numbers of shipments for personal computers are not shown in Figs. 3.4 and 3.5. But the desktop personal computers' shipments declined from 5,192,000 in 2006 fiscal year (April 2006–March 2007) to 1,753,000 in 2015 fiscal year (April 2015–March 2016). Similarly, the notebook personal computers' shipments declined from 6,858,000 in 2006 fiscal year (April 2006–March 2007) to 53,582,015 fiscal year (April 2015–March 2016) (Japan Electronics and Information Technology Industries Association 2017).

Collected personal computers are treated and recycled at very high rates (over 70%). But notebook personal computers collected from the households are recycled at the lower rates around 57.1% (in 2015 fiscal year (April 2015–March 2016) according to the Personal Computer Reduce, Reuse and Recycle (often termed as "PC3R") Promotion Association (2017). The Personal Computer Reduce, Reuse and Recycle Promotion Association states that after the collected 12.87million units of personal computers go through the final treatments, 2.8 million units will be reused domestically, 3.64 million units will be recycled as resources domestically, 0.35 million units will be put in the landfills, 2.15 million units will be exported overseas for reuse purpose, and 3.08 million units will be set aside for resource purposes in the 2014 fiscal year (April 2014–March 2015).

Table 3.5 Electronic waste and recycled appliances

Panel A: Yields (metals, plastics, etc.) derived from collected small appliances		
Item		**Weight (tons)**
Total small electric appliances collected		57,260
Metals recycled after processing		36,567[a]
Recycled plastics		2550
Incinerated plastics		13,612
Reused appliances		149
Residuals		4298
Panel B: The estimated amounts of electronic-waste generated by small appliances		
Item	**Number of recycled units**	**Weight (tons)**
Electric shaver, jar, electric pots, and others	61,368,572	185,179
Mobile phones, phones, and others	47,842,169	16,053
Speaker (automobile), digital camera, DVD	90,400,559	132,750
PC, printer, monitor and others	22,868,114	140,290
Bulbs, electric lighting equipment	795,062,951	110,055
Camera	91,057	37
Clock	82,431,127	12,384
Desktop game machine, portable game machine	13,223,334	12,916
Electric calculator, digital dictionary	10,273,500	1129
Electronic thermometer, sphygmomanometer etc.	22,229,256	20,576
Electronic keyboard, electric guitar, etc.	1,089,299	4459
The handheld game machine, and mobile toys	1,128,449	186
Electric drill and others	6,633,000	14,100
AC adapter, controller and others	2,109,710	427
Total	**1,156,751,096**	**650,539**

Source: compiled by the authors using information for public use from the Japanese Ministry of Environment (2012)

[a]Breakdown of various metals collected out of the total recycled metals in tons:

iron	26,326;	stainless steel and brass	148;
aluminum	2023;	gold, silver and palladium	2798;
copper	1469;	other metals	6573.

Although we don't have a detailed breakdown of the recycled metals for mobile phones, we have statistics for small electric appliances including mobile phones. We summarize these as follows for 2015 fiscal year (April 2015–March 2016) (Table 3.5).

Table 3.6 shows that the recycling and processing of Electronic-waste from small appliances, personal computers, and mobile phones give potentially significant amounts of valuable metals. This observation recently prompted the Tokyo Organising Committee of the 2020 Olympic and Paralympic Games to decide that Olympic medals for winners of the Tokyo games are to be made of metals distilled from mobile phones and called for the local governments and the local post offices in Japan to collect them for recycling.

Table 3.6 Metals included in the used small electric and electronic appliances

Metals (tons)	Total: small appliances	Mobile phones	PCs
Iron (Fe)	230,105	418	16,845
Aluminum (Al)	24,708	50	3914
Copper (Cu)	22,789	1001	2730
Lead (Pb)	740	19	220
Zinc (Zn)	649	44	70
Silver (Ag)	68.9	10.5	21.1
Gold (Au)	10.6	1.9	4.5
Antimony (Sb)	117.5	2.3	43.5
Tantalum (Ta)	33.8	3.2	14.9
Tungsten (W)	33.0	27.1	1.1
Neodymium (Nd)	26.4	18.9	–
Cobalt (Co)	7.5	2.2	–
Bismuth (Bi)	6.0	0.7	0.8
Palladium (Pd)	4.0	0.5	2.1

Source: Compiled by the authors using information for public use from the Japanese Ministry of Environment (2012)

3.3.1 Supply Chain Implications of Recycling End Products: Reductions of the Resources Used in Upstream Suppliers

We know that all electric and electronic appliances we consider here are manufactured products whose production processes consist of many stages of inputs from the upstream suppliers. Many of these upstream inputs are basic and precious metals which remain in the final products as the output from the relevant supply chains. For this reason, it is important to look at the behavior of not only the final product Electronic-waste but also many inputs of electronic nature (electronic commodities) that were used in the upstream production processes of the supply chains. For these reasons, we consider Electronic-wastes as consisting of toxic and nontoxic wastes generated throughout the upstream stages by suppliers of the supply chain.

For example, the first and third panels of Table 3.7, respectively, show the amounts of industrial wastes that are generated by one million yen worth of production of personal computers and one million yen worth of production of mobile phones. These wastes generated consist of 37 types of industrial wastes in all self- and other industry sectors that form the upstream stages of the supply chains. Table 3.3 shows the waste outputs for the top five industry sectors, as well as the total amounts of industrial wastes generated for each electronic product. The amounts that were landfilled of the final wastes generated after the recycling and process treatments are presented in Table 3.8. For example, the first panel of Table 3.8 shows that after recycling and processing of Electronic-wastes, one million yen worth of personal computer production generated 7.8 kg of residual to be landfilled.

Table 3.7 Industrial wastes, directly and indirectly, generated from the unit production of e-commodities

	I-O industry sector specific for the product	
	1 million yen worth of PC production	Induced wastes (kg)
Industrial wastes generated in	Electricity	23.1
	Other electronic components	22.9
	Pig iron	20.7
	Paper	10.1
	Printing, plate making, and bookbinding	8.3
	Total	156.2
	1 million yen worth of electronic computing equipment (accessory equipment) production	Induced wastes (kg)
Industrial wastes generated in	Electronic computing equipment (accessory equipment)	65.9
	Electricity	28.5
	Pig iron	26.4
	Other electronic components	21.6
	Crude steel (converters)	10.6
	Total	247.6
	1 million yen worth of cellular phones production	Induced wastes (kg)
Industrial wastes generated in	Pig iron	28.2
	Electricity	26.8
	Other electronic components	22.5
	Copper	17.0
	Plastic products	14.5
	Total	218.7
	1 million yen worth of electric measuring instruments production	Induced wastes (kg)
Industrial wastes generated in	Pig iron	35.9
	Other electronic components	25.4
	Electric measuring instruments	24.2
	Electricity	23.1
	Crude steel (converters)	14.8
	Total	201.5
	1 million yen worth of liquid crystal element production	Induced wastes (kg)
Industrial wastes generated in	Liquid crystal element	48.4
	Electricity	42.0
	Pig iron	16.2
	Other electronic components	10.2
	Printing, plate making, and bookbinding	8.8
	Total	216.2

Source: By authors' calculations. The methodology and the data used are presented in Hayami et al. (2015). Further details are also given in Asakura et al. (2001), Hayami and Nakamura (2007), Hayami et al. (2008), and Japanese Ministry of Internal Affairs and Communications (2017)

Table 3.8 Amounts landfilled: induced (directly and indirectly) wastes after treatments of recycled Electronic-wastes

	1 million yen worth of PC production	Induced wastes after treatment: landfilled (kg)
Industrial wastes generated in	Printing, plate making, and bookbinding	1.3
	Paper	0.9
	Other nonferrous metals	0.8
	Electricity	0.8
	Copper	0.5
	Total	7.8
	1 million yen worth of electronic computing equipment (accessory equipment) production	Induced wastes after treatment: landfilled (kg)
Industrial wastes generated in	Printing, plate making, and bookbinding	1.1
	Electronic computing equipment (accessory equipment)	1.0
	Electricity	0.9
	Other nonferrous metals	0.8
	Paper	0.7
	Total	9.3
	1 million yen worth of cellular phone production	Induced wastes after treatment: landfilled (kg)
Industrial wastes generated in	Copper	2.2
	Printing, plate making, and bookbinding	1.3
	Paper	1.0
	Lead and zinc (inc. regenerated lead)	0.9
	Plastic products	0.9
	Total	11.4
	1 million yen worth of electric measuring instrument production	Induced wastes after treatment: landfilled (kg)
Industrial wastes generated in	Other nonferrous metals	1.0
	Printing, plate making, and bookbinding	1.0
	Electricity	0.8
	Paper	0.7
	Copper	0.7
	Total	8.6
	1 million yen worth of liquid crystal element production	Induced wastes after treatment: landfilled (kg)
Industrial wastes generated in	Other nonferrous metals	1.4
	Electricity	1.4
	Printing, plate making, and bookbinding	1.4
	Liquid crystal element	0.8
	Paper	0.7
	Total	10.0

Source: By authors' calculations. The methodology and the data used are presented in Hayami et al. (2015). Further details are also given in Asakura et al. (2001), Hayami and Nakamura (2007), Hayami et al. (2008), and Japanese Ministry of Internal Affairs and Communications (2017)

3.3.2 Reductions in Emissions of Greenhouse Gases from Recycling Electronic-Waste

Greenhouse gases are typically measured in terms of carbon dioxide (often noted as "CO_2") equivalent in tons. Greenhouse gases are by-products of most production processes along the stages of supply chains where production inputs such as electricity and metals are used by the suppliers. So we can analyze possible reductions in the generation of greenhouse gases resulting from the recycling of Electronic-waste. Analysis of the implications of Electronic-waste recycling for reductions in greenhouse gas emissions along the supply chains can be done by employing the input-output method, using the input-output tables and some relevant data of the kinds we used in the previous Sect. 3.3.1. To save space we only present certain summary results illustrating how the recycling of Electronic-waste could potentially reduce emissions of greenhouse gases and hence contribute to the possible solution to the global warming problem.

We are particularly interested in measuring the impact of the following government policy-driven form of recycling of Electronic-waste on the reductions in the generation of greenhouse gas emissions (measured in carbon dioxide equivalent measured in tons) generated from the production processes. The particular government policy of our interest to analyze here is called the Eco Policy, which gives some (not insignificant) rewards to the users of older-generation home appliances if they recycle them and buy newer more energy-efficient appliances with equivalent functions. (See, e.g., Japanese Ministry of Environment (2011), Japan Environmental Management Association for Industry (2013, 2017), and Hotta et al. (2014), for details of this policy.)

Rewards are given in terms of some level of subsidy for the purchase of newer-generation appliances. Based on the actual implementation of this social experiment, many (but not all) consumers owning older energy-inefficient appliances had chosen to recycle their old appliances and buy newer appliances using the rewards. The Eco Policy was implemented for a limited period of May 2009–March 2011. The appliances covered in this program are air conditioners, refrigerators, and television sets. The analysis reported in the Japanese Ministry of Environment (2011) divides the periods of analysis into three time periods: May 2009–March 2010, April 2010–December 2010, and January 2011–March 2011. Given the initial consumers who own particular appliances, some fractions of them choose to replace their old products with new ones. They benefit from the Eco Policy and their information is shown under the "replacement purchase." Some consumers who did not own particular appliances may choose to buy new units. Their information is shown under "new unit purchase." In the last period, January 2011–March 2011, only "new unit purchase" occurs since replacements are no longer allowed under the Eco Policy. Using carbon dioxide emission rates and the power consumption rates estimated elsewhere for older and newer appliances, the difference in the amount of greenhouse gas emissions that were saved by consumers' purchases of newer appliances is calculated and shown in Table 3.9. The total reductions in carbon dioxide

Table 3.9 Reductions in greenhouse gas emissions (CO_2 equivalent): Japan's Eco Policy experiments (2009–2011)

	Number of units	Replacements purchased (%) (rate of recycling)	Replacements purchased (number)	Year purchased of old unit	Electric power consumed old units/year (KWH/year)	Electric power consumed new units/year (KWH/year)	Reductions (%) in electric power used	Estimated total CO_2 emissions (tons/KWH)	Reductions in CO_2 emissions (tons/year)	Total reduction in CO_2 emissions (tons)
May 2009–March 2010 (replacement purchase)										
Air conditioner	2,668,000	47.6%	1,269,968	95	1396	1138	258 (18%)	0.000561	183,813	May 2009–March 2010: 948045 tons
Refrigerator	2,838,000	70.6	2,003,628	95	822	343	479 (58)	0.000561	538,413	
Television set	14,347,000	65.5	9,397,285	98	151	122	29 (19)	0.000561	152,884	
TOTAL	19,853,000		12,670,881						875,110	
May 2009–March 2010 (new unit purchase)										
Air conditioner	2,668,000	52,4%	1,398,032		1193	1138	55 (5%)	0.000561	43,136	
Refrigerator	2,838,000	29.4	834,372		377	343	34 (9)	0.000561	15,915	
Television set	14,347,000	34.5	4,949,715		127	122	5 (4)	0.000561	13,884	
TOTAL	19,853,000		7,182,119						72,935	
April 2010–December 2010 (replacement purchase)										

Air conditioner	6,507,000	44.8%	2,915,136	96	1373	1046	328 (24%)	0.000561	536,408	April 2010–December 2010: 1587688 tons
Refrigerator	3,529,000	73.1	2,579,699	96	783	318	465 (59)	0.000561	672,953	
Television set	20,185,000	68.3	13,786,355	99	143	96	47 (33)	0.000561	363,505	
TOTAL	30,221,000		19,281,190						1,572,866	
April 2010–December 2010 (new unit purchase)										
Air conditioner	6,507,000	55.2%	3,591,864		1048	1045	3 (05)	0.000561	6045	
Refrigerator	3,529,000	26.9	949,301		321	318	3 (1)	0.000561	1598	
Television set	20,185,000	31.7	6,398,645		98	96	2 (2)	0.000561	7179	
TOTAL	30,221,000		10,939,810						14,822	
January 2011–March 2011 (new unit purchase only)										
Air conditioner	233,000	45.6%	106,248	96	1373	973	400 (29%)	0.000561	23,842	January 2011–March 2011: 1782041 tons
Refrigerator	272,000	72.0	195,840	96	783	270	513 (66)	0.000561	56,361	
Television set	5,054,000	67.1	3,391,234	99	143	83	60 (42)	0.000561	114,149	
TOTAL	5,559,000		1,693,322						194,352	
May 2009–March 2011 total reductions in CO_2 emissions due to the Eco Policy: 4317774 tons										

Source: Compiled by the authors using information for public use from the Japanese Ministry of Environment (2011) and JLCA (2013)

equivalent emissions for the three time periods are estimated to be 4,317,774 tons. These reductions are deemed to be the effects of the Eco Policy for the three appliances.

3.3.3 Issues of Who Bears the Burden of the Costs of Electronic-Waste Recycling

We have pointed out above that the recycling rates for waste electrical and electronic equipment are generally low (mostly under 30% of recycling ready Electronic-waste in Japan; see Table 3.2). We have also pointed out that the costs associated with recycling Electronic-waste, including the recycling fees as well as the costs of transportation and waste material removal, which are to a large extent borne by the consumer, are relatively high in general. Japanese policy discussions on Electronic-waste recycling have also raised the issues related to the lack of transparency in the determination of the Electronic-waste recycling cost (Recycling Working Group 2007).

Electronic-Waste and European Union

Unlike the four categories of Electronic-waste considered by Japanese laws (i.e., air conditioners, television sets, electric refrigerators and freezers, and electric washing machines and clothes dryers), European Union's Waste Electrical and Electronic Equipment Directive specifies the following ten Electronic-waste categories: (1) large household appliances; (2) small household appliances; (3) information technology equipment; (4) consumer equipment (television sets, etc.); (5) lighting appliances; (6) power tools; (7) toys, leisure, and sports equipment; (8) medical equipment; (9) monitoring and control instruments; and (10) vending machines and automatic teller machines.

Another area of European Union's Electronic-waste management that differs from Japanese practices is in the areas of allocation of the responsibility for collection and the allocation of costs. For example, producers are responsible for their own new products, but that all producers shall cover costs jointly when products that are already on the market are discarded by consumers. Until 2011 (2013 for large white goods), however, producers will be permitted to add waste processing costs to the prices of new products separately (visible fee).

In general, the European Union regulations differ from Japan's in a number of ways. The European Union laws cover a broad range of products, assign responsibility and costs to producers, establish collection targets and recycling rates, and limit the use of hazardous substances (Yoshida and Yoshida 2010). As OKOPOL (2007) notes, European Union's policy aim is to build a system that, by these means, recovers waste electrical and electronic equipment separately rather than disposing of it as municipal solid waste. We note that the municipality is an important stakeholder in the European Union's waste electrical and electronic equipment recycling system.

In terms of performance, the European Union's overall collection rate, however, is not so high compared to Japan's. Based on Table 3.10, Yoshida and Yoshida (2010) observe that although, in 2005, with a per capita recovery amount of 5.13 kg (Japan's per capita collection amount for the four types is 3.5 kg), the European Union had more than attained its 4-kg target; nevertheless, between individual countries, considerable differences remain: Sweden had collected 12.20 kg and the United Kingdom 9.9 kg, whereas the Czech Republic in Eastern Europe had recovered only 0.33 kg (Table 3.1). A look at the collection rates for each of the ten categories shows that, among the ten product categories, refrigerators and air conditioners account for 27% of the possible total, with 40% for large household appliances, 28% for information technology equipment, 30% for cathode-ray tube CRT TVs, and 65% for monitoring and control instruments (United Nations University and AEA Technology 2007, Table 56). According to the recent Waste Electrical and Electronic Equipment Forum data for 2007, the per capita recovery amount is nearly 7.80 kg, and 11 countries have collectively managed to collect over 4.0 kg.

According to estimation by Makela (2009), of the amounts of waste electrical and electronic equipment the European Union had collected for treatment, only 33% of them were treated and 13% were landfilled properly within EU; while 54%

Table 3.10 Collection performance in the European Union countries and Japan by category, 2005

Category number totals											
Country	1	2	3	4	5	6	7	8	9	10	1–10
Japan	2.58	n.d.	n.d.	0.82	n.d.	n.d.	n.d.	n.d.	n.d.	n.d.	NA
Norway	8.15	0.46	2.68	2.01	–	–	0.04	0.06	–	0.01	13.41
Switzerland	4.19	1.40	3.52	2.17	0.12	0.04	0.01	0.00	0.00	0.00	11.44
Austria	2.00	0.3	0.1	0.2	0.1	Inc 2	Inc 2	Inc 2	Inc 2	Inc 2	2.77
Belgium	2.99	1.12	1.16	1.64	0.20	0.14	0.00	0.02	0.00	0.00	7.26
Czech Republic	0.14	0.00	0.12	0.05	0.00	0.00	0.00	0.01	0.00	0.01	0.33
Estonia	0.48	0.00	0.04	0.10	n.d.	n.d.	n.d.	n.d.	n.d.	n.d.	0.63
Finland	4.75	0.28	1.44	1.30	0.27	0.03	0.00	0.02	0.01	0.00	8.10
Hungary	0.91	0.04	0.09	0.22	0.01	0.00	0.00	0.00	0.00	0.00	1.27
Ireland	6.68	0.28	0.43	0.67	0.09	0.07	n.d.	n.d.	0.00	n.d.	8.22
Netherlands	2.59	0.53	n.d.	1.18	0.03	0.06	0.03	0.00	0.00	0.02	4.44
Slovakia	0.35	0.04	0.05	0.20	0.02	0.00	0.00	0.00	0.00	0.00	0.66
Sweden	5.01	1.41	2.54	2.36	0.74	0.11	0.02	0.02	n.d.	n.d.	12.20
UK	7.17	0.54	0.59	1.10	0.04	0.35	0.16	0.00	0.00	0.00	9.95
I=NO=CH average	4.97	0.93	3.10	1.67	0.06	0.02	0.02	0.03	0.00	0.01	10.80
Euro average	3.11	0.42	0.65	0.88	0.14	0.08	0.02	0.01	0.00	0.00	5.31

Figures here show the amounts of wastes collected per inhabitant (kg) by country and category. See the source for details on the used categories (1–10). Categories 1–10 represent the total aggregates
Notes: *n.d.* no data, *Inc 2* included in category 2 figure, *NA* figure not available
Source: United Nations University and AEA Technology (2007), Table 43, page 74: Collection performance (kg per inhabitant) by Category

of them were submitted to substandard treatment both inside and outside the European Union.

3.4 Concluding Remarks

In this chapter, we have considered a variety of factors, including legal, statistical, economic, and organizational factors, that affect the recycling of waste electrical and electronic equipment or more broadly the recycling of general Electronic-wastes in Japan and other countries.

Despite significant efforts on the part of governments at all levels as well as other stakeholders, collecting, recycling, and processing of Electronic-waste remain to be a difficult task. Generally, there is a consensus that Electronic-waste continues to increase as per capita gross domestic product increases. Such increases are significant not only in developed nations but also in developing nations as well. This necessarily implies that shipping out Electronic-waste out of developed countries to developing countries is no longer a viable means of Electronic-waste disposal.

These substances that make up the Electronic-waste contain valuable resources, some of which are toxic and cannot be simply put away for landfills. We summarize our findings on the aspects of production and waste management systems, broadly termed environmental management systems, that need to be redesigned in an integrated manner.

1. Design products so as to minimize Electronic-waste while their functions remain intact. Also, design products so that the ultimate recycling of the products could be done with ease.
2. Design recycling-related facilities that can efficiently recycle process valuable metal and other resources.
3. Design a system that allocates the responsibilities among the stakeholders of Electronic-waste recycling policies based on their respective incentives.
4. Estimating accurately the relevant costs and benefits of alternative methods of collecting, recycling, and processing of Electronic-waste is important. Then, how such costs and benefits are to be allocated among the stakeholders consistent with their respective economic incentives is also important.
5. Manufacturing supply chains must be taken into account when Electronic-waste recycling and processing policies are formulated. It is important to clarify which upstream suppliers are responsible for particular components of Electronic-waste.
6. Another important policy issue is to decide what the primary objectives of Electronic-waste recycling are for the nation. Is it to reduce the greenhouse gas emissions that are emitted in the production process by minimizing the use of Electronic-waste causing metals and other materials? Or is it simply to reduce the amounts that go to landfills?

Environmental management of Electronic-waste requires consideration of many of these and other issues by the government policymakers as well as other stakeholders, including consumers, producers, and other private sector and public sector parties.

References

Asakura K, Hayami H, Mizoshita M, Nakano S, Nakamura M, Shinozaki M, Washizu A, Yoshioka K (2001) The input-output table for environmental analysis (Kankyo Bunsekiyo Sangyo Renkanhyo in Japanese). Keio University Press, Tokyo

Baldé CP, Wang F, Kuehr R, Huisman J (2015) The global electronic waste monitor 2014: quantities, flows and resources. United Nations University, Bonn

Enechange (2016) Recycling costs of home appliances (in Japanese). Tokyo. https://enechange.jp/articles/kaden-rycycle-cost-list. Accessed 16 June 2018

EU Eurostat (2017) Waste statistics-electrical and electronic equipment. November 2017. http://ec.europa.eu/eurostat/statistics-explained/index.php/Waste_statistics_-_electrical_and_electronic_equipment. Accessed 16 June 2018

Hayami H, Nakamura M (2007) Greenhouse gas emissions in Canada and Japan: sector-specific estimates and managerial and economic implications. J Environ Manag 85(2):371–392

Hayami H, Nakano S, Nakamura M, Suzuki M (2008) The input-output table for environmental analysis and applications (Kankyo Bunsekiyo Sangyo Renkanhyo to sono Oyo, in Japanese). Keio University Press, Tokyo

Hayami H, Nakamura M, Nakamura A (2015) Economic performance and supply chains: the impact of upstream firms' waste output on downstream firms' performance in Japan. Int J Prod Econ 160:47–65

Hotta Y, Santo A, Tasaki T (2014) EPR-based electronic home appliance recycling system under home appliance recycling act of Japan. Tokyo. https://www.oecd.org/environment/waste/EPR_Japan_HomeAppliance.pdf. Accessed 16 June 2018

Japan Electronics and Information Technology Industries Association (JEITA) (2017) Shipment of cellular phones. Tokyo. https://www.jeita.or.jp/japanese/stat/cellular/2017/index.htm. Accessed 16 June 2018

Japan Environmental Management Association for Industry (JEMAI) (2013, 2017) Recycle data book. Tokyo. http://www.cjc.or.jp/data/databook.html. Accessed 16 June 2018

Japan Times (2015, May 9) Electronic waste recycling still falling short. Tokyo

Japanese Ministry of Economy, Trade and Industry (METI) (2017) FY 2016 enforcement status of the home appliances recycling law and recycling statistics for manufacturers and importers. Tokyo

Japanese Ministry of Environment (2001) Laws: waste and recycling. Available from: http://www.env.go.jp/en/laws/recycle/06.pdf. Accessed 17 Mar 2019

Japanese Ministry of Environment (2011). Eco policy and effects, (in Japanese). Available from: http://www.env.go.jp/policy/ep_kaden/pdf/effect.pdf

Japanese Ministry of Environment (2012) Central environment committee 31 January 2012. Available from: http://www.env.go.jp/press/files/jp/19123.pdf. Accessed 17 Mar 2019.

Japanese Ministry of Internal Affairs and Communications (2017) Input-output table for Japan. Tokyo. http://www.soumu.go.jp/english/dgpp_ss/data/io/index.htm. Accessed 15 Feb 2018

Japanese Ministry of Justice (2017) Tokyo. http://www.japaneselawtranslation.go.jp/?re=02. Accessed 15 Feb 2018

Kusch S, Hills CD (2017) The link between electronic waste and gross domestic product: new insights from data from the Pan-European region. Resources 6(2):15

LCA Society of Japan (JLCA) (2013) LCA database. Life Cycle Assessment Society of Japan. http://lca-forum.org/database. Accessed 16 June 2018

Makela T (2009) The waste electrical and electronic equipment directive-business generation. In: Proceedings of IERC 2009, Salzburg, pp 183–190

Mobile Recycle Network (MRN), and also Japan Environmental Management Association for Industry (JEMAI) (2018). http://www.mobile-recycle.net. Accessed 18 June 2018

OKOPOL (Okopol GmbH Institute for Environmental Strategies), IIIEE (The International Institute for Industrial Environmental Economics, Lund University), and RPA (Risk & Policy Analysts) (2007) The producer responsibility principle of the waste electrical and electronic equipment directive, final report. Hamburg

PC3R Promotion Association, Tokyo (2017). http://www.pc3r.jp. Accessed 18 June 2018

Recycling Working Group, Industrial Structure Council Environment Subcommittee, Japanese Ministry of Economy, Trade and Industry (2007) Lowering and making transparent of the recycling prices and recycling costs (in Japanese). Tokyo. http://www.meti.go.jp/committee/gizi_1/14.html#meti0003770. Accessed 18 June 2018

United Nations University and AEA Technology (2007) 2008 Review of Directive 2002/96 on Waste electrical and electronic equipment (WEEE), final report, 05 August 2007, United Nations University, Bonn

Yoshida F, Yoshida H (2010) Japan, the European Union, and waste electronic and electrical equipment recycling: key lessons learned. Environ Eng Sci 27(1):21–28

Chapter 4
Recent Technologies in Electronic-Waste Management

Mohamed Aboughaly and Hossam A. Gabbar

Contents

Abstract The electrical and electronic industry generates more than 50 million metric tonnes of Electronic-waste annually from discarded and obsolete equipment. According to the Environmental Protection Agency (EPA), 7 million tonnes of electronic equipment become obsolete each year, making Electronic-waste the most rapidly growing waste stream in the world. Electronic-waste often contains hazardous materials as well as base metals such as zinc, copper and iron that can reach up to 60.2% in Electronic-waste products such as refrigerators, washing machines and TVs. Global legislation and regulations play an important role in Electronic-waste recycling strategies and cover 66% of electronic industry practices; most importantly to be mentioned are waste electrical and electronic equipment (WEEE)

M. Aboughaly (✉) · H. A. Gabbar
Faculty of Energy Systems and Nuclear Science, and Faculty of Engineering and Applied Science, University of Ontario Institute of Technology, Oshawa, ON, Canada
e-mail: mohamed.aboughaly@uoit.net; hossam.gabbar@uoit.ca

A. Khan et al. (eds.), *E-waste Recycling and Management*,
Environmental Chemistry for a Sustainable World 33,
https://doi.org/10.1007/978-3-030-14184-4_4

directive, restriction of hazardous substances (RoHS) directive and registration, evaluation, authorization and restriction of chemicals (REACH) directive regulations.

Waste electrical and electronic equipment (WEEE) are classified into four categories which are photovoltaic (PV) panels, cathode ray tube (CRT), liquid crystal displays (LCDs) and light-emitting diode (LED) displays, computers and laptops and cell phones. Photovoltaic panels are a common silicon-based electronic equipment with 65% recycling rate. The recycling process starts with glass and aluminium recovery followed by thermal treatment at 650° C. Another category is liquid crystal displays and light-emitting diode displays which consume 70% of global indium production, while its recycling requires manual sorting and separation, solvent extraction and acid leaching, respectively. Additionally, cell phones have the lowest recycling rate due to the complexity of recycling caused by compact design and high production rate. Lithium is considered the most valuable recycling material in cell phones and smart batteries. In terms of viable Electronic-waste thermal treatment, thermal plasma consumes 2 kWh/kg in both pyrometallurgical and hydrometallurgical recycling processes. It plays an important role in the recovery of heavy metals such as silver, gold, lead and copper due to high energy density, gas flux temperature and ionization that increases reactivity.

Keywords Electronic-waste management · Contaminants · Material composition · Electronic-waste regulations · Waste generation · Metal recovery · Metallurgical recycling · Thermoplastics in Electronic-waste · Hazards in electronic recycling · Recycling hierarchy

Abbreviations

ABS	Acrylonitrile butadiene styrene
Cd	Cadmium
Cu	Copper
DC	Direct current
ELV	End-of-life vehicles directive
EOL	End-of-life electronics
EPA	Environmental Protection Agency, USA
GDP	Global domestic product
HIPS	High impact polystyrene
Ni	Nickel
Pb	Lead
PC	Polycarbonates
PPO	Polyphenylene oxide
REACH	Registration, evaluation, authorization and restriction of chemicals
RF	Radio frequency

RoHS	Restriction of hazardous substances directive
Sn	Tin
WEEE	Waste electrical and electronic equipment
Zn	Zinc

4.1 Introduction and Overview

Waste electrical and electronic equipment (WEEE) is defined as obsolete equipment made of electric circuits and electromagnetic fields, which are discarded due to inability to deliver expected performance or after the end of its lifetime (Tanskanen 2013). This chapter discusses end-of-life recycling challenges, influences waste management strategies implemented in industrialized nations to avoid environmental emissions and utilizes useful heavy metals and recyclable components. Electrical and electronic equipment (EEE) is increasingly growing in volume and material diversity and considered one of the largest growing global manufacturing activities (Buekens and Yang 2014). Electronic-waste is in tremendous global increase with more than 67.7 million mobile phones, Internet access to more than 50% of global households, 91% households that own televisions, and 99% of households that have refrigeration (Tuba 2015). The Electronic-waste main source is obsolete televisions, computers, mobile phones, printers, electronics, home appliances and equipment used in the manufacturing and automotive industry (Sthiannopkaoa and Wong 2013). On the other hand, the rapid social and economic development of Electronic-waste management is evolving as a major concern particularly in industrial and developing nations (Mukhtar et al. 2016).

Recycling requires sustainable and Electronic-waste management which is rapidly growing to approximately 40 million tonnes in 2014 (Cao et al. 2016). The exponential increase in Electronic-waste strives in the manufacturing industry and governments to draw attention to electrical and electronics material recycling for better management of renewable and non-renewable resources. Below is an ordinary recycling and separation strategy commonly applied in Electronic-waste recycling facilities as shown in Fig. 4.1 (Hai et al. 2017):

The most common thermoplastics of high interest in waste electrical and electronic equipment recycling are acrylonitrile butadiene styrene (ABS), high impact polystyrene (HIPS), polyphenylene oxide (PPO), polycarbonate (PC), polyvinyl chloride (PVC) and polypropylene (PP) which retain a high production value. Below is the electrical and Electronic-waste plastic composition in computer manufacturing facilities as shown in Fig. 4.2 (Hester and Harrison 2009):

- A sharp increase of high-impact polystyrene (HIPS) in televisions and screen panels, stereos and fans
- A sharp increase of polypropylene (PP) in fans
- A sharp increase of poly-methyl-methacrylate (PMMA) in vacuum cleaners
- A sharp increase of acrylonitrile butadiene styrene (ABS) in computers and vacuum cleaners

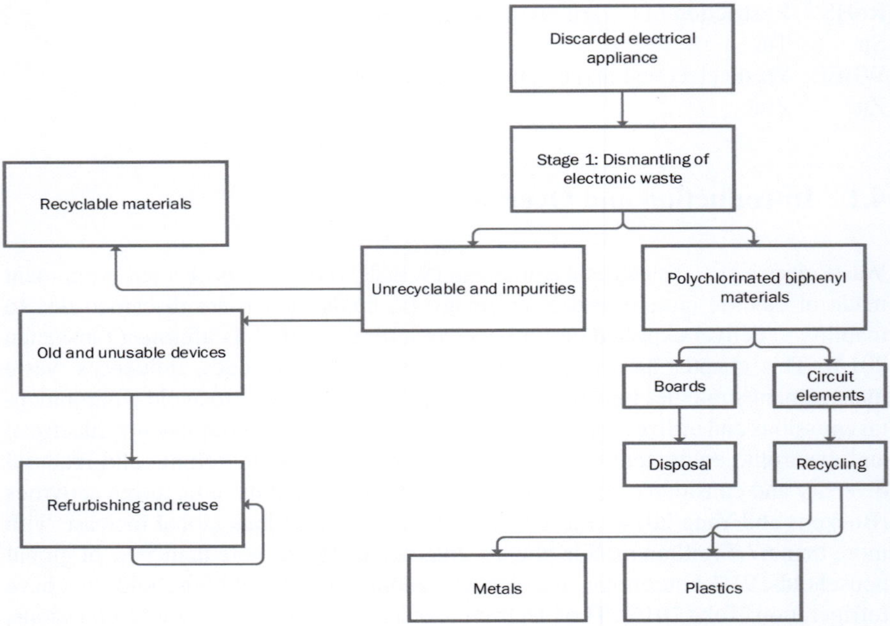

Fig. 4.1 Schematic diagram of global Electronic-waste separation and recycling strategies. (Hai et al. 2017)

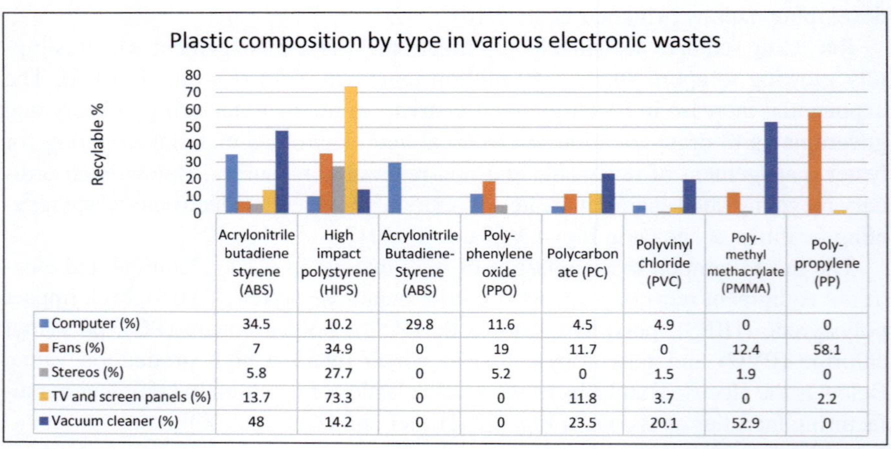

	Acrylonitrile butadiene styrene (ABS)	High impact polystyrene (HIPS)	Acrylonitrile Butadiene-Styrene (ABS)	Poly-phenylene oxide (PPO)	Polycarbon ate (PC)	Polyvinyl chloride (PVC)	Poly-methyl methacrylate (PMMA)	Poly-propylene (PP)
Computer (%)	34.5	10.2	29.8	11.6	4.5	4.9	0	0
Fans (%)	7	34.9	0	19	11.7	0	12.4	58.1
Stereos (%)	5.8	27.7	0	5.2	0	1.5	1.9	0
TV and screen panels (%)	13.7	73.3	0	0	11.8	3.7	0	2.2
Vacuum cleaner (%)	48	14.2	0	0	23.5	20.1	52.9	0

Fig. 4.2 Plastic composition by type in various Electronic-wastes. (Hester and Harrison 2009)

4.1.1 The Problem Scale

Due to rapid development in industrial nations that produce a growing amount of Electronic-waste where the majority of Electronic-waste is produced in North America and East Asia, respectively (Veit and Bernardes 2015a), sustainable management of Electronic-waste is considered one of the major challenges in the twenty-first century, and it is found that countries with the highest gross domestic product (GDP) produce more Electronic-waste due to higher industrial waste generation causing an exponential increase over time (Kusch and Hills 2017). The existence of expensive metals in the Electronic-waste stream can retain a major economic benefit for recycling industries if appropriate techniques are used. Added to that, recycling of Electronic-waste prevents hazardous materials from existing in landfills which reduces harm to soil and habitats (Seeberger et al. 2016). The current estimation of global Electronic-waste produced is approximately 41 million tonnes at an annual increase of 3–5% per year (Kumar et al. 2017). Also, Electronic-waste legislations cover 66% of our globe but are still considered ineffective. Thus, more accurate Electronic-waste policies are required for more reliable recycling, data collection and waste management. Below are two major environmental concerns in recycling facilities for Electronic-waste feedstocks, which recommend pretreatments before recycling (Kiddee et al. 2013):

- Soil contamination caused by toxic substances from non-recycled disposals and primitive recycling processes
- Contamination of hazardous materials such as lead, mercury, arsenic, copper, barium and chromium which have a direct impact on workers and employees who do labour in E-waste recycling areas causing health problems

Hazardous materials in Electronic-waste affect human health and habitats during disposals. Metals can reach up to 60.2% of Electronic-waste and electronic equipment such as refrigerators, washing machines, air conditioners and televisions which are considered as the main sources of base metals such as aluminium, copper, lead and zinc (Veit and Bernardes 2015b). Another major problem in Electronic-waste is a generation of toxic fumes in the uncontrolled environment due to burning or recycling activities caused by toxic materials. This includes iron, copper, aluminium and gold which form around 60% of Electronic-waste (Widmer et al. 2005). Also, an increase in demand by communities and improvement of people's lives has made the electrical and electronic industry the centre of technology and attention in modern world economies leading to more waste being generated (Öztürk 2015). Below is a flowchart of main biological and environmental concerns from different categories of Electronic-waste products as shown in Fig. 4.3:

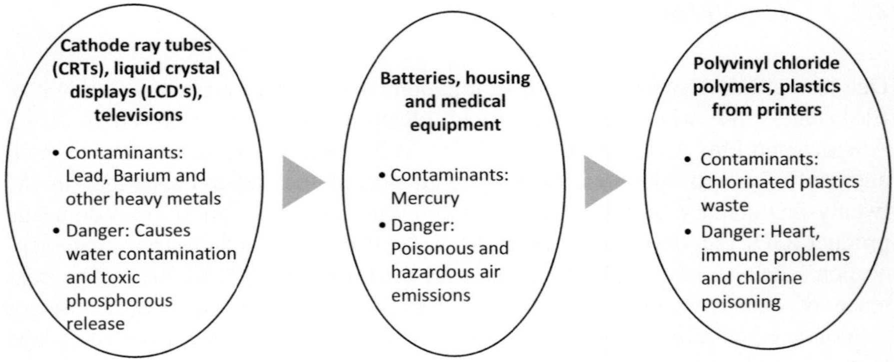

Fig. 4.3 Contaminants in Electronic-waste products

Added to that, below are major challenges faced in Electronic-waste recycling facilities:

- *High Electronic-waste volume*: High electronic volumes are generated due to rapid growth in the electronic industry and high demand for new technologies which are expected to peak by 2030 (van Santen et al. 2010). These high waste volumes increase the cost of disposal and storage and encourage waste generators to seek recycling alternatives such as incineration and landfilling which increases the environmental impact.
- *Poor design and complex integrated Electronic-waste system*: Electronic-waste complex designs impose many challenges including mixing of different non-recyclable materials as well as bolts and screws that increases the complexity of the recycling process in compact products such as mobile phones and liquid crystal display (LCD) screens. This causes recycling to be more intensive and sophisticated which reduces process efficiency and increases cost.
- *Lack of Electronic-waste management and recycling regulations*: Lack of effective enforcement of accurate E-waste regulations both in management and recycling. Many geographical areas lack adequate and accurate regulations, which cause ineffective procedures of E-waste management.

4.1.2 Collection Practices of Electronic-Waste

The European Union (EU) has the highest rates of Electronic-waste recycling estimated at 60–80% of global waste followed by Japan (Zimring and Rathje 2012). However, much of the Electronic-waste is discarded in general waste streams or illegally exported to giant importers of Electronic-waste from Asian nations such as

China, India and Pakistan. These illegal recycling practices are motivated by capturing of valuable substances such as copper, iron, silicon, nickel and gold recycling (Zhang et al. 2012). Currently, approximately 70% of the world's global Electronic-waste is available in China at approximately 28 million tonnes of Electronic-waste per annum (Perkins et al. 2014). Below are recommendations for effective industrial E-waste recycling strategies:

- Effective analysis of Electronic-waste generation including economic and population data to establish a correlation between various factors such as population density with waste generation and waste recycling per capita (Kumar et al. 2017).
- Analysis of future trend of Electronic-waste systems by the electronic and electrical equipment sales trends and estimated life of electrical and electronic products.
- Increase the awareness and understanding of recycling including analysis of Electronic-waste and the importance of recycled materials present in Electronic-waste as well as preserving the environment from hazardous metallic and non-metallic content.
- Implementation of streamline software and usage of material tracking in production chains for identification of bottleneck in the production chains.
- High-level waste collection recovery implementation in provinces and the prevention of waste exportation to developing countries (Kumar et al. 2017).
- Obtaining accurate waste generation and recycling data considering hidden business information by exporters and importers (Tran and Salhofer 2018).
- Accurate classification of global Electronic-waste recovery regions as shown in Fig. 4.4 as well as the development of Electronic-waste hazardous materials regulations, which includes 1000 different substances. It is to be noted that more than 70% of heavy metals in US landfills come from Electronic-wastes (Widmer et al. 2005).

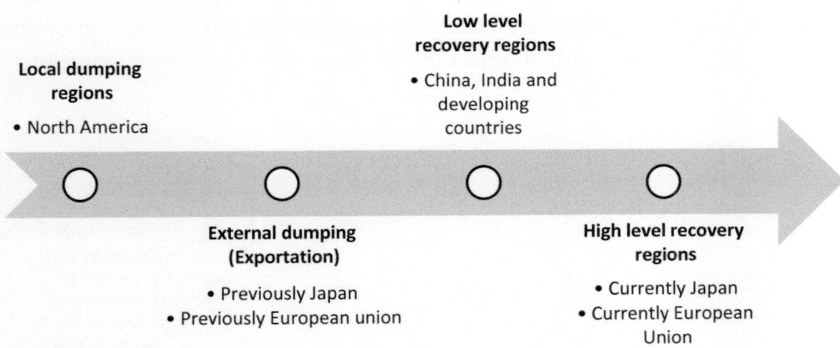

Fig. 4.4 Global Electronic-waste recovery regions. (Singh et al. 2016)

4.1.3 Challenges and Influences on Electronic Recycling

The electronic recycling industry faces challenges during different industrial stages such as collection, chemical analysis, transportation and electronic recycling facilities as follows:

- Smaller wearable electronics are difficult to recycle due to the difficulty of disassembly. This can be solved at the design stage to avoid shredding at the end-of-life processing stage (Ceballos and Dong 2016).
- The high cost of chemical analysis to identify materials in recycling feedstock, which limits extraction of all useful recyclable materials.
- Lack of environmental regulations to protect the health of employees during recycling.
- Updating outdated occupational exposure limits for hazardous chemicals present in recycled electronics.
- Poor emission control in recycling facilities due to the usage of old equipment and emissions from pyrometallurgical processes (Ceballos and Dong 2016).

4.1.4 The Recycling Hierarchy and Recyclable Markets

According to the Environmental Protection Agency (EPA), the Electronic-waste hierarchy is a process in series, which includes reduction, reuse and recovery of Electronic-waste materials as shown in Fig. 4.5 (Parajuly and Wenzel 2017). The most preferred Electronic-waste practice is reduction through durable designs and usage of recyclable contents in new products (Matsumoto et al. 2016). Electronic-waste recycling contains many heavy metals that require special end-of-life handling such as cadmium (Cd), nickel (Ni), copper (Cu), lead (Pb) and thermoplastics (Zhang et al. 2015a). Advantages of Electronic-waste recycling implementation include minimization of pollution, reduction of landfilling space and prevention of long-term damage to soil and habitats as well as avoidance of Electronic-waste

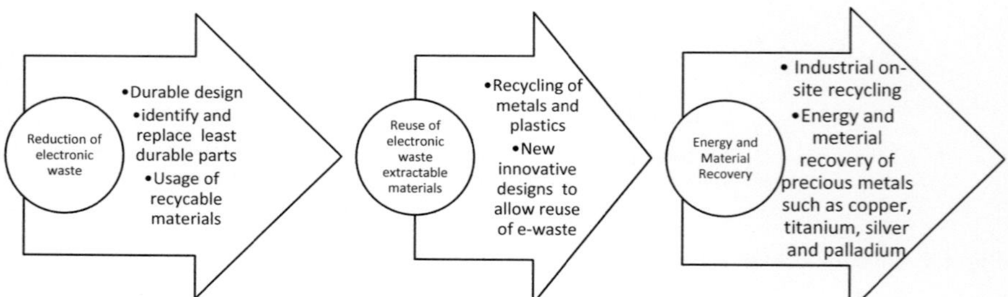

Fig. 4.5 Hierarchy of recycling strategies of Electronic-waste. (Parajuly and Wenzel 2017)

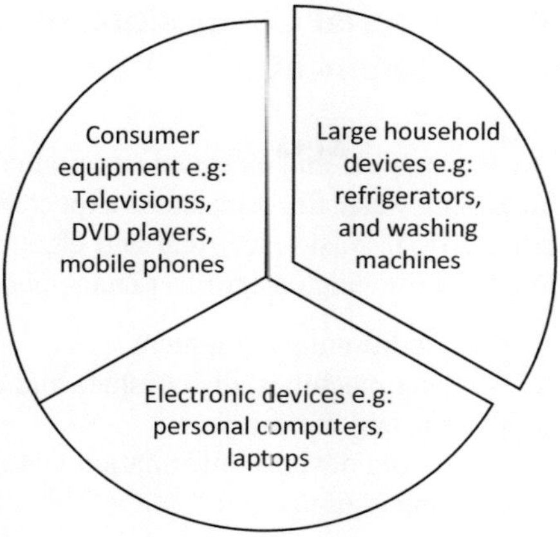

Fig. 4.6 Discarded Electronic-waste main categories at recycling facilities. (Perkins et al. 2014)

exportation. It is approximated that than 80% of 50 million tonnes of E-waste generated from developed countries is shipped as E-waste export to developing countries (Perkins et al. 2014).

Also, Electronic-waste generated from households is divided into three categories as shown in Fig. 4.6. Nearly 70% of deposited global E-waste is unreported or arrives from unknown sources, which shows the importance of E-waste recycling regulations. (Ongondo et al. 2011).

4.1.5 *Waste Electric and Electronic Management and Control Regulations*

Although many countries have their own management and control E-waste regulations, most developing nations rely on developed nations management and control regulations (Awasthi and Li 2017). The Environmental Protection Agency (EPA) is considered one of the most famous organizations that states laws and regulations for electrical and Electronic-waste manufacturing, Electronic-waste handling and recycling (Veit and Bernardes 2015a). The European Union (EU) imposes main laws for electronics manufacturers through the collection, recycling and deposition of E-waste through the following directives (Bai et al. 2016):

- Restriction of the use of certain hazardous substances (RoHS)
- Waste electrical and electronic equipment (WEEE) regulations
- End-of life-vehicle (ELV) regulations (European Communities 2000)

As shown above, all manufacturers have to comply with Environmental Protection Agency laws and regulations regarding their internal material and product regulations during manufacturing, reuse, repair, recycling and disposal of waste electrical and electronic equipment.

4.2 Material Composition in Waste Electrical and Electronic Equipment

Waste electrical and electronic equipment (WEEE) is considered one of the fastest increasing waste deposits and is expected to reach 50 million tonnes by 2018 (Balde et al. 2015). Waste electrical and electronic equipment (WEEE) can be classified into the following categories (Zhang and Xu 2016):

1. Air-conditioning equipment
2. Washing machines, dishwashers and cooking devices
3. TV sets and monitors
4. Household devices, information technology and telecommunication equipment
5. Lighting equipment

The essential features of retrieving components from Electronic-waste during recycling involve three industrial stages as shown below (Fig. 4.7):

4.3 Current and Future Electronic-Waste Management Technologies

The science and government communities believe that improved waste management achieves better economic, environmental and social benefits. This can be achieved through better management and separation of waste streams as well as better estimations of waste compositions (Cucchiella et al. 2015). Below is a classification of selected waste stream materials that accounts for large volumes in electrical and Electronic-waste:

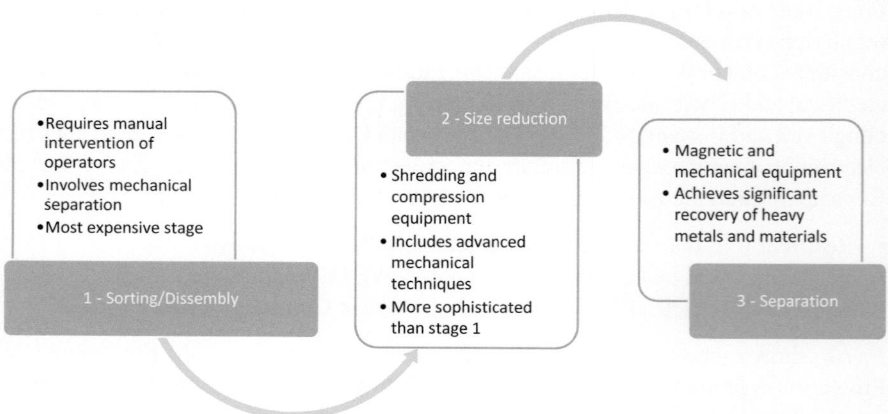

Fig. 4.7 Industrial stages for component separation in waste electrical and electronic equipment (WEEE). (Dalrymple et al. 2007)

4.3.1 Photovoltaic Panels

Photovoltaic panels are considered common electric equipment with an expected lifespan of 20 years that consists of 90% silicon-based components as well as the small percentage of key metals such as cadmium, tellurium, indium and gallium (Drouiche et al. 2014). The economic advantages are never obtained by recyclers and manufacturers, which motivates them to perform landfilling and incineration practices. Development of more advanced recycling techniques helps to increase recycling up to more than 90%. Both current collection and recycling rates of PV panels are up to 65% in 2016 (Granata et al. 2014). Also, carbon emissions and energy cost of recycling photovoltaic panels are relatively low in comparison with other waste electrical and electronic materials. Most common materials in photovoltaic panels are crystalline silicon either monocrystalline or polycrystalline, cadmium telluride (CdTe), amorphous silicon or copper-indium selenide (CIS) (Granata et al. 2014). Below are the operational stages of photovoltaic panels recycling as shown in Fig. 4.8:

4.3.2 Cathode Ray Tube Displays, Liquid Crystal Displays and Light-Emitting Diode Displays and Monitors

Liquid crystal display (LCD) has replaced cathode ray tube display (CRT) in the previous two decades and is considered one of the most important electronic equipment that accounts for 80% of computer and electronic sales (Bhakar et al. 2015). These monitors contain benzene, cyano-groups, chlorine and other toxic substances as well as precious heavy metals such as gold (Au), silver (Ag) and lead (Pb) in printed wiring board (PWB) that requires careful handling during Electronic-waste management. The most valuable material from liquid crystal display screens is

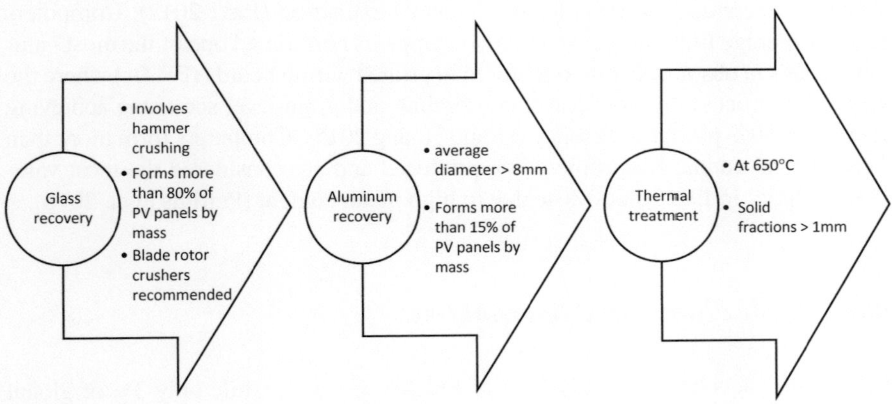

Fig. 4.8 Operation stages of photovoltaic panels industrial recycling

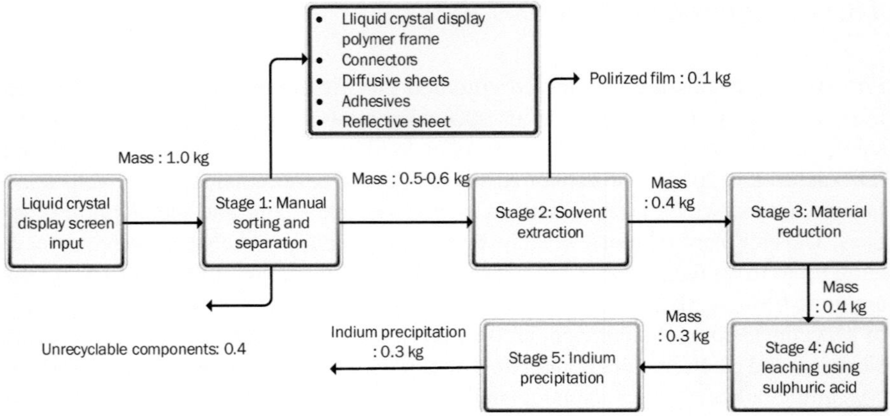

Fig. 4.9 Mass balance of indium recovery from liquid crystal displays

indium due to its scarcity and high consumption of liquid crystal display production which account for 70% of produced indium worldwide (Zhang et al. 2015b).

Most common recycling technique from waste liquid crystal display recycling includes dismantling which involves the removal of hazardous materials followed by removal of indium and valuable materials. Below is a recommended industrial flowchart for separation of indium as a precipitate from liquid crystal display panels as shown in Fig. 4.9 (Ruan et al. 2012):

4.3.3 Computers, Notebooks and Laptops

Advancement in the Electronic-waste handling of computers shows that masses of computers reduce with time which leads to a reduction of computer Electronic-waste but increases the complexity of metals extraction (Ravi 2012). Computers have an average lifespan of 3 years. Also, copper is considered one of the most valuable metals in obsolete computers stored in printed wiring boards (PWBs) where the separation process involves crushing, grinding and grain-level screening achieving more than 80% process efficiency (Li and Huang 2015). Computers form more than 38% of total annual Electronic-waste produced and are considered the most valuable category in Electronic-waste due to high metal content (Petridis et al. 2016).

4.3.4 Cell Phones and Smartphones

Cell phones account for nearly 17% of Electronic-waste, while only 3% of global mobile phones are recycled (Li et al. 2017). Lithium ions are considered the most valuable material available in batteries of smartphones (Li et al. 2014). Cell phones

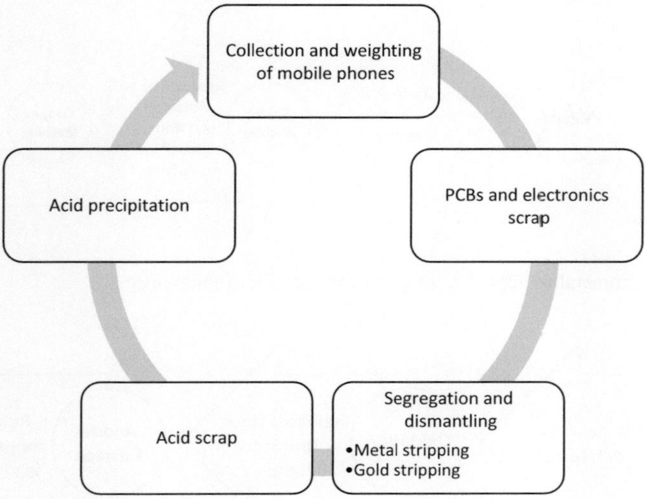

Fig. 4.10 Mobile phones industrial recycling cycle. (Kaya 2016)

have a relatively short lifespan, and recyclers face problems due to compact design and waste released to the environment during recycling (Soo and Doolan 2014). Below is a flowchart of mobiles recycling as shown in Fig. 4.10. Also, below are important related to mobile phones waste management (Sarath et al. 2015):

- Mobile phones waste generation statistics
- Consumer behaviour studies
- Economics of mobile phones recycling stages
- Toxicity assessment of mobile phone parts including new materials replacement assessment
- Materials identification and recovery methods

4.3.5 Thermal Plasma Technology in Electronic-Waste Recycling

The high temperature of thermal plasma and high energy density are utilized in solid waste treatments for more than a decade where high carbon content waste is converted to syngas and undesired products such as oxides or slag (Mitrasinovic et al. 2011). For Electronic-waste, thermal plasma is utilized in the extraction of valuable heavy metals such as (Ag) silver, gold (Au), Lead (Pd) and copper (Cu) (Khaliq et al. 2014). Since E-waste is classified as a hazardous material, different process routes are used to extract metals using pyrometallurgical, electrometallurgical and hydrometallurgical processes. Below is a flowchart of chemical processes used for metals separation for hydrometallurgical and pyrometallurgical processes (Figs. 4.11 and 4.12, respectively) (Torres and Lapidus 2016):

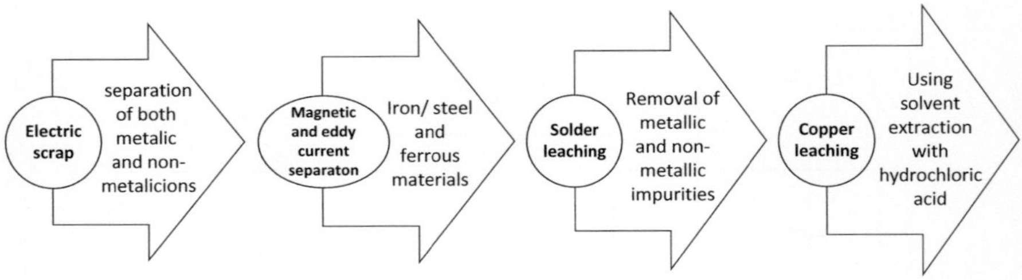

Fig. 4.11 Hydrometallurgical recycling processes of Electronic-waste

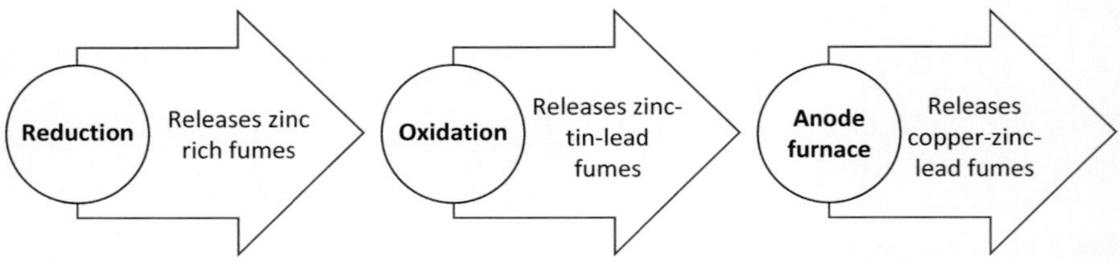

Fig. 4.12 Pyrometallurgical recycling processes of Electronic-waste

Thermal plasma acts as an excellent heating source for pyrometallurgical processes due to its high energy density, the presence of ions that increases reactivity and high ion/plasma jet gas temperature that justifies its excellent thermal profile (Rath et al. 2012). Direct current (DC) or radio-frequency (RF) thermal plasma usually requires 1400–1600° C for melting of metals after separation of non-metallic impurities. At an earlier stage, metals should be separated from plastics and non-metallic impurities in a mechanical disassembly process before shredded metals (i.e. particle size: 10–15 cm) are deposited in a plasma reactor. The average energy consumption using thermal plasma for E-waste treatment is 2 kWh/kg (Rath et al. 2012). The thermal plasma separates metals such as copper, iron, aluminium and nickel. Below is a flowchart of thermal plasma pyrometallurgical processes (Fig. 4.13):

4.4 Conclusion

This chapter defines and classifies Electronic-waste separation and recycling strategies as well as mentioning the importance of Electronic-waste management and statistics of the exponential increase of Electronic-waste. Also, electronic waste is classified into three main categories, and valuable metals and contaminants are extracted in each category. In addition, major challenges faced in Electronic-waste recycling include a high volume of Electronic-waste due to the rapid growth of the electronic industry and high demand of Electronic-waste management which is

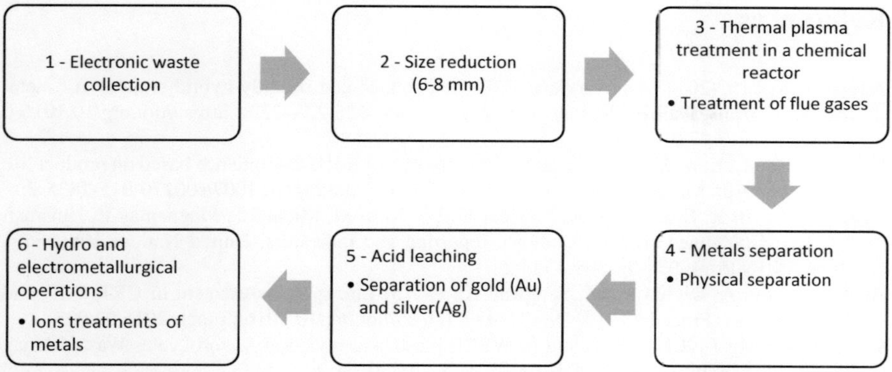

Fig. 4.13 Thermal plasma treatment pyrometallurgical processes

expected to reach $50 billion by 2020. Moreover, poor design and complexity of electronics impose difficulties during recycling and separation of different materials, which can be overcome by alternative designs and in-house industrial recycling. Also, lack of Electronic-waste management regulations inhibits effective recycling and allows E-waste producers to practice landfilling and incineration.

Also, Electronic-waste major categories include photovoltaic panels which have the longest lifespan of electronic products of 20 years and consist of 90% silicon-based products. Advantages of photovoltaic (PV) panels recycling include low-carbon emissions and low energy costs in comparison with other major electronic products. Also, liquid crystal display (LCD) panels contain toxic substances that require pretreatment before recycling of heavy metals especially indium which is considered the most valuable material in liquid crystal display (LCD) screens. On the other hand, computer E-waste focuses on copper extraction from printed wiring boards (PWBs) which have a process efficiency of 80%. On the other hand, mobile phones have very low recycling rate that doesn't exceed 3% where lithium is the most valuable material available in batteries. Since mobile phones have a short lifespan and a compact design, difficulties are faced in segregation and dismantling as well as the acid scrapping of metals.

Thermal plasma is used in both hydrometallurgical and pyrometallurgical chemical processes for the extraction of heavy metals due to its high energy density and capability of increasing reactivity. It is utilized to extract metals such as silver and gold at 1400–1600° C. Thermal plasma reactors require pre-shredding of metals to a particle size of 10–15 cm and an energy intensity of 2 KWh/kg.

Acknowledgements Authors would like to thank Mr. Janak Handa – ProFlange, GPROSYS team and Advanced Plasma Engineering Lab (APEL), ESCL (Energy Safety and Control Lab) at UOIT for their support to this research work.

References

Awasthi AK, Li J (2017) An overview of the potential of eco-friendly hybrid strategy for metal recycling from WEEE. Resour Conserv Recycl 126:228–229. https://doi.org/10.1016/j.resconrec.2017.07.014

Bai CZW, Gao L, Kang S (2016) Optimization research of RoHS compliance based on product life cycle. Int J Adv Manuf Technol 84:1777–1785. https://doi.org/10.1007/s00170-015-7835-2

Balde CP, Kuehr R, Blumenthal K, Fondeur Gill S, Kern M, Micheli P, Magpantay E, Huisman J (2015) Guidelines on classifications, reporting and indicators. United Nations University, Bonn. ISBN Print: 978-92-808-4553-2

Bhakar V, Agur A, Digalwar AK, Sangwan KS (2015) Life cycle assessment of CRT, LCD and LED monitors. Procedia CIRP 29:432–437. https://doi.org/10.1016/j.procir.2015.02.003

Buekens A, Yang J (2014) Recycling of WEEE plastics: a review. J Mater Cycles Waste Manag 16(3):415–434. https://doi.org/10.1007/s10163-014-0241-2

Cao J, Chen Y, Shi B, Lu B, Zhang X, Ye X, Zhai G, Zhu C, Zhou G (2016) WEEE recycling in Zhejiang province, China: generation, treatment, and public awareness. J Clean Prod 127:311–324. https://doi.org/10.1016/j.jclepro.2016.03.147

Ceballos DM, Dong Z (2016) The formal electronic recycling industry: challenges and opportunities in occupational and environmental health research. Environ Int 95:157–166. https://doi.org/10.1016/j.envint.2016.07.010

Cucchiella F, D'Adamo I, Koh SCL, Rosa P (2015) Recycling of WEEEs: an economic assessment of present and future e-waste streams. Renew Sust Energy Rev 51:263–272. https://doi.org/10.1016/j.rser.2015.06.010

Dalrymple I, Wright N, Kellner R, Bains N, Geraghty K, Goosey M, Lightfoot L (2007) An integrated approach to electronic waste (WEEE) recycling. Circuit World 33(2). https://doi.org/10.1108/03056120710750256

Drouiche N, Cuellar P, Kerkar F, Medjahed S, Boutouchent-Guerfi N, Hamou MO (2014) Recovery of solar grade silicon from kerf loss slurry waste. Renew Sust Energ Rev 32:936–943. https://doi.org/10.1016/j.rser.2014.01.059

European Communities (2000) Directive 2000/53/EC of the European Parliament and of the Council of 18 September 2000 on end-of life vehicles. Off J Eur Commun 269:34–42. doi: 2004R0726 - v.7 of 05.06.2013

Granata G, Pagnanelli F, Moscardini E, Havlik T, Toro L (2014) Recycling of photovoltaic panels by physical operations. Sol Energy Mater Sol Cells 123:239–248. https://doi.org/10.1016/j.solmat.2014.01.012

Hai HT, Hung HV, Quang ND (2017) An overview of electronic waste recycling in Vietnam. J Mater Cycles Waste Manage 19:536–544. https://doi.org/10.1007/s10163-015-0448-x

Hester RE, Harrison RM (2009) Electronic waste management, design and analysis application, vol 27. RSC Publishing, Cambridge. https://doi.org/10.1007/s10163-015-0448-x

Kaya M (2016) Recovery of metals from electronic waste by physical and chemical recycling processes. Int J Chem Mol Nucl Mater Metall Eng 10:259–270. https://doi.org/10.1016/j.wasman.2016.08.004

Khaliq A, Rhamdhani MA, Brooks G, Masood S (2014) Metal extraction processes for electronic waste and existing industrial routes: a review and Australian perspective. Resources 3(1):152–179. https://doi.org/10.3390/resources3010152

Kiddee P, Naidu R, Wong MH (2013) Electronic waste management approaches: an overview. Waste Manage 33:1237–1250. https://doi.org/10.1016/j.wasman.2013.01.006

Kumar A, Holuszko M, Espinosa DCR (2017) E-waste: an overview on generation, collection, legislation and recycling practices. Resour Conserv Recycl 122:32–42. https://doi.org/10.1016/j.resconrec.2017.01.018

Kusch S, Hills CD (2017) The link between E-waste and GDP new insights from data from the Pan-European region. Resources 6(15):1–10. https://doi.org/10.3390/resources6020015

Li X, Huang W (2015) Process for copper recovery from e-waste: printed wiring boards in obsolete computers. Nat Environ Pollut Technol 14:145–148

Li L, Zhai L, Zhang X, Lu J, Chen R, Wu F, Amine K (2014) Recovery of valuable metals from spent lithium-ion batteries by ultrasonic-assisted leaching process. J Power Sources 262:380–385. https://doi.org/10.1016/j.jpowsour.2014.04.013

Li J, Ge Z, Liang C, An N (2017) Present status of recycling waste mobile phones in China: a review. Environ Sci Pollut Res 24:16578–16591. https://doi.org/10.1007/s11356-017-9089-z

Matsumoto M, Masui K, Fukushige S, Kondoh S (2016) Sustainability through innovation in product life cycle design. Springer, Singapore. https://doi.org/10.1007/978-981-10-0471-1

Mitrasinovic A, Pershin L, Wen JZ, Mostaghimi J (2011) Recovery of Cu and valuable metals from e-waste using thermal plasma treatment. J Miner Met Mater Soc 63:24–28. https://doi.org/10.1007/s11837-01

Mukhtar EM, Williams ID, Shaw PJ, Ongondo FO (2016) A tale of two cities: the emergence of urban waste systems in a developed and a developing city. Recycling 1:254–270. https://doi.org/10.3390/recycling1020254

Ongondo FO, Williams ID, Cherrett TJ (2011) How are WEEE doing? A global review of the management of electrical and electronic wastes. Waste Manag 31(4):714–730. https://doi.org/10.1016/j.wasman.2010.10.023

Öztürk T (2015) Generation and management of electrical–electronic waste (e-waste) in Turkey. J Mater Cycles Waste Manag 17:411–421. https://doi.org/10.1007/s10163-014-0258-6

Parajuly K, Wenzel H (2017) Product family approach in e-waste management: a conceptual framework for circular economy. Sustainability 9(768):1–14. https://doi.org/10.3390/su9050768

Perkins DN, Drisse M-NB, Nxele T, Sly PD (2014) E-waste: a global hazard. Ann Glob Health 80(4):286–295. https://doi.org/10.1016/j.aogh.2014.10.001

Petridis NE, Stiakakis E, Petridis K, Dey P (2016) Estimation of computer waste quantities using forecasting techniques. J Clean Prod 112:3072–3085. https://doi.org/10.1016/j.jclepro.2015.09.119

Rath SS, Nayak P, Mukherjee PS, Chaudhury GR, Mishra BK (2012) Treatment of electronic waste to recover metal values using thermal plasma coupled with acid leaching – a response surface modeling approach. Waste Manag 32:575–583. https://doi.org/10.1016/j.wasman.2011.11.001

Ravi V (2012) Evaluating overall quality of recycling of e-waste from end-of-life computers. J Clean Prod 20:145–151. https://doi.org/10.1016/j.jclepro.2011.08.003

Ruan J, Guo Y, Qiao Q (2012) Recovery of indium from scrap TFT-LCDs by solvent extraction. Proc Environ Sci 16:545–551. https://doi.org/10.1016/j.proenv.2012.10.075

Sarath P, Bonda S, Mohanty S, Nayak SK (2015) Mobile phone waste management and recycling: views and trends. Waste Manag 46:536–545. https://doi.org/10.1016/j.wasman.2015.09.013

Seeberger J, Grandi R, Kim SS, Mase WA, Reponen T, Ho S-M, Chen A (2016) E-waste management in the United States and public health implications. J Environ Health 79(3):8–16

Singh N, Li J, Zeng X (2016) Global responses for recycling waste CRTs in e-waste. Waste Maange 57:187–197. https://doi.org/10.1016/j.wasman.2016.03.013

Soo VK, Doolan M (2014) Recycling mobile phone impact on life cycle assessment. Procedia CIRP 15:263–271. https://doi.org/10.1016/j.procir.2014.06.005. 21st CIRP Conference on Life Cycle Engineering

Sthiannopkaoa S, Wong MH (2013) Handling e-waste in developed and developing countries: initiatives, practises, and consequences. Sci Total Environ 463–464:1147–1153. https://doi.org/10.1016/j.scitotenv.2012.06.088

Tanskanen P (2013) Management and recycling of electronic waste. Acta Mater 61:1001–1011. https://doi.org/10.1016/j.actamat.2012.11.005

Torres R, Lapidus GT (2016) Copper leaching from electronic waste for the improvement of gold recycling. Waste Manag 57:131–139. https://doi.org/10.1016/j.wasman.2016.03.010

Tran CD, Salhofer SP (2018) Analysis of recycling structures for e-waste in Vietnam. J Mater Cycles Waste Manag 20:110–126. https://doi.org/10.1007/s10163-016-0549-1

Tuba O (2015) Generation and management of electrical-electronic waste (e-waste) in Turkey. J Mater Cycles Waste Manag 17(3):411–421. https://doi.org/10.1007/s10163-014-0258-6
van Santen RA, Khoe D, Vermeer B (2010) 2030: technology that will change the world. Oxford University Press, New York. https://doi.org/10.1002/anie.201106671
Veit HM, Bernardes AM (2015a) Electronic waste: generation and management. Springer, Cham. https://doi.org/10.1007/978-3-319-57076-1_12
Veit HM, Bernardes AM (2015b) Electronic waste: recycling techniques. Springer, Cham. https://doi.org/10.1007/978-3-319-15714-6
Widmer R, Oswald-Krapf H, Sinha-Khetriwal D, Schnellmann M, Böni H (2005) Global perspectives on e-waste. Environ Impact Asses Rev 25:436–458. https://doi.org/10.1016/j.eiar.2005.04.001
Zhang L, Xu Z (2016) A review of current progress of recycling technologies for metals from waste electrical and electronic equipment. J Clean Prod 127(20):19–36. https://doi.org/10.1016/j.jclepro.2016.04.004
Zhang K, Schnoor JL, Zeng EY (2012) E-waste recycling: where does it go from here? Environ Sci Technol 2(46):10861–10867. https://doi.org/10.1021/es303166s
Zhang Z, Yu M-L, Zhang J-H, Wang X, Jiang J-H (2015a) Distribution characteristics of heavy metals in E-waste recycling sites. Nat Environ Pollut Technol 14:137–140
Zhang K, Wu Y, Wang W, Li B, Zhang Y, Zuo T (2015b) Recycling indium from waste LCDs: a review. Resour Conserv Recycl 104:276–290. https://doi.org/10.1016/j.resconrec.2015.07.015
Zimring CA, Rathje WL (2012) Encyclopedia of consumption and waste: the social science of garbage. SAGE, Los Angeles. https://doi.org/10.4135/9781452218526

Chapter 5
Recycling Challenges for Electronic Consumer Products to E-Waste: A Developing Countries' Perspective

Patricia Guarnieri, Lúcio Camara e Silva, Lúcia Helena Xavier, and Gisele Lorena Diniz Chaves

Contents

P. Guarnieri (✉)
Faculty of Economics, Administration, Accounting and Public Policies Management, University of Brasilia, Brasilia, Brazil

L. Camara e Silva
Management Engineering Department, CDSID – Center for Decision Systems and Information Development, Federal University of Pernambuco, Recife, Brazil
e-mail: lucio@cdsid.org.br

L. H. Xavier
Centre for Mineral Technology (CETEM), Rio de Janeiro, Brazil
e-mail: lxavier@cetem.gov.br

G. L. D. Chaves
Department of Engineering and Technology, Federal University of Espírito Santo, Vitória, Espírito Santo, Brazil
e-mail: giselechaves@ufes.br

A. Khan et al. (eds.), *E-waste Recycling and Management*, Environmental Chemistry for a Sustainable World 33,
https://doi.org/10.1007/978-3-030-14184-4_5

Abstract Recycling and sustainable development issues are increasing in importance around the world. This aspect is more prominent in developing countries, in which there are many informal recycling activities and few environmental legislations regulating waste management. This chapter discusses the recycling challenges regarding the adoption of e-waste reverse logistics under the perspective of developing countries. For this purpose, we gathered information from papers published in international databases and reports such as the United Nations Environment Programme and Global e-waste Monitor, thus identifying data available to American countries (Brazil, Argentina, Chile and Mexico), South Africa and Asian countries (China, India, Russia, Indonesia, Turkey, Pakistan, South Korea, Thailand and Singapore). As key findings we can point out the categorization of the barriers into financial/economics, environmental, market related, legal, policy related, management, knowledge related and technical and technological related. As main contributions of this chapter, we can highlight (i) the compilation of information related to recycling challenges of e-waste in developing countries, and (ii) the identification of some solutions and actions to overcome these barriers is also performed, which can be useful for practitioners and researchers in this field.

Keywords Circular economy · Developing countries · E-waste · Electronic residues · Reconditioning · Recycling challenges · Reverse logistics · Residues revalorization · Urban mining · Waste management

5.1 Introduction

It can be observed that recycling and the related issue of sustainable development are increasing in importance around the world (de Oliveira et al. 2012). Although reverse logistics and waste management practices are in state of infancy in developing countries representing a great challenge (Sasaki and Araki 2013; Abdulrahman et al. 2014; Bouzon et al. 2015; Ferri et al. 2015), many legislations focused on waste management have arisen over the past decades (Kumar et al. 2017; Nnorom and Osibanjo 2008; Appelbaum 2002).

A valuable category of waste and the most rising one in the world (Awasthi et al. 2018) is the electric and electronic equipments waste, or E-waste. Based on the literature, we can highlight the evaluation of the assessment of E-waste management in European Union (Appelbaum 2002), North America and developing and emerging countries (Ongondo et al. 2011). However, there are different patterns of E-waste management in general guided by regulations or economic reasons. Despite the federal E-waste regulation in force in Brazil, China and India, many problems related to E-waste hazardous substances and occupational health can be observed in those countries. In this context, it is worthful to analyse the behaviour of E-waste management in the world to provide suitable policies, strategies and sustainable solutions for efficient decision making.

A sensible aspect that impacts E-waste management is the economic value of the materials found in waste electrical and electronic equipment, such as valuable metals as copper, gold and silver, as well as critical materials as tungsten, niobium and cobalt. Thus, it must be considered the hazardous potential in its handling – a real question to be faced by stakeholders and government when the workforce infrastructure is poorly automated.

However, in the early 1990s, many organizations have been conceived to deal with E-waste secondary products, since the waste electrical and electronic equipment (E-waste) and Restriction of Certain Hazardous Substances directives have been published. This type of organizations is focused on the circular business model and has the challenge of accomplishing sustainability in a *stricto sensu* way. Thus, over the last two decades, huge countries were established to recover strategic and critical elements from E-waste, and it may be the most up-to-date case in waste management history. Regarding 'sustainability', Awasthi et al. (2018) state that it is an ideal and a moral compromise requiring a social commitment in waste management, especially in developing countries where the waste pickers are acting instead of specific machinery.

Recycling tools for waste management, in general, seem to be provided according to demand and technical resources available, and closing loop is the implementation of a best practice long-term orientation. However, to invest in primitive E-waste recycling is a hazardous alternative for health and environment quality (Jiang et al. 2017).

Some critical barriers to E-waste management must be faced by stakeholders to meet sustainable requirements. In this context, the aim of this chapter is to analyse the main recycling challenges related to E-waste management in developing countries. For this purpose, the main barriers to E-waste reverse logistics adoption in developing countries were also discussed, and some solutions were indicated.

As main findings of this research, we can point out the categorization of the barriers into financial/economics, environmental, market related, legal, policy related, management, knowledge related and technical and technological related. Finally, as main contributions of this chapter are the compilation of information related to recycling challenges of E-waste in developing countries, and the identification of some solutions and actions to overcome these barriers is also performed, which can be useful for practitioners and researchers.

5.2 Theoretical Background

5.2.1 Reverse Logistics of e-waste

The definition of reverse logistics comprises different approaches presented in regulations, standards or technical studies around the world. The proposal of the American Reverse Logistics Executive Council (Rogers and Tibben-Lembke 1998) considers the steps of logistics management in reverse order (from consumption to the origin) and adds the appropriate destination and the recapturing value as main purposes. Nonetheless, the value maximization, logistics steps and product life cycle are emphasized by Guide and Van Wassenhove (2009) reverse logistics' definition.

Reverse logistics concept was previously presented as the take-back system adopted in the environmental European directives, in which each state member developed the specific regulation. Some examples of these European directives are the End of Life Vehicles and Directive 2000/53/European Community, the Battery Directive 2006/66/European Community, Waste Electrical and Electronic Equipment Directive 2012/96/European Community and Restriction of Certain Hazardous Substances Directive 2011/65/European Community. This procedure was carried on for e-waste and other categories of post-consumer products.

Nowadays, the reverse logistics concept merged the closed-loop supply chain proposal (Govindan and Soleimani 2017) and can even be interpreted as one of the tools of circular economy concept. Likewise, the concept of urban mining emerges as a derivation of the proposed circularity. This concept contributes by reducing the exploitation of natural resources and prioritizing the use and reuse of products and materials. The urban mining works on as an alternative to obtaining inputs (precious metals) from secondary resources from E-waste.

The most recent database on E-waste was published in Global E-waste Monitor (Baldé et al. 2017). This report states that in 2016 Asia produced over 18 metric tonnes of E-waste, followed by Europe (12.3 metric tonnes), the Americas (11.3 metric tonnes), Africa (2.2 metric tonnes) and Oceania (0.7 metric tonnes). Therefore, some attention needs to be direct to provide priority to environmental regulation in Asian countries. In fact, developing countries generally lack actions to regulate and to prevent negative environmental impacts resulting from E-waste handling (Kumar et al. 2017).

In addition, Table 5.1 presents the main studies related to reverse logistics and management of E-waste in developing countries.

5.2.2 *Barriers to Implementation of Reverse Logistics of Waste Electrical and Electronic Equipment (E-waste)*

While Achillas et al. (2010) have attempted to emphasize the importance to consider the development of a reverse network, several studies deal on barriers to implementing reverse logistics in developed countries. In general, the barriers for E-waste recycling alternatives have its origin in different aspects, such as economic, political or technical issues, and are managed according to the particularities of each country.

Abdulrahman et al. (2014) studied the barriers of reverse logistics in China and found four categories: (i) management, (ii) financial, (iii) policy and (iv) infrastruc-

Table 5.1 Main studies related to reverse logistics and management of E-waste in developing countries

Country	Main aspects highlighted	Authors
China	Informal E-waste management associated with serious environmental and health impacts, the supply deficiency of formal recyclers and the safety problems of remanufactured electronic products	Chi et al. (2014)
Turkey	This study provides an analysis of the level of consumer choice of E-waste in Nigde-Aksaray Province. A large-scale recycling scheme with economic incentives and accommodations must be developed for Turkey's E-wastes	Gök et al. (2017)
Turkey	This study proposes a reverse logistics system for E-waste in Turkey that was simulated with ten scenarios. Although it is benefited from the generic models in the literature, the proposed mathematical model mainly differs from the existing models by considering recycling rates for each product category	Kilic et al. (2015)
Turkey	Based on the literature review and the gathered information from the electrical and electronic equipment's manufacturers and the municipalities interviewed, a conceptual E-waste management model for Turkey was built	Camgöz-Akdag and Aksoy (2014)
Turkey	This study explains the technological familiarity of Turkey's population and their technological product possession together with related E-waste production potential and calculates the estimated E-waste amounts raised by years	Öztürk (2015)
Thailand	The development of E-waste forecasting model based on surveyed data is developed and utilized in predicting the amount of E-waste in Thailand	Chirapat et al. (2012)
Pakistan	Evidence of the e-waste scenario in Pakistan indicating the major E-waste recycling sites, current and future domestic generation of E-waste, hidden flows or import of E-waste and discusses various challenges for E-waste management	Iqbal et al. (2015)
Brazil	This study approaches the problem of structuration of reverse logistics of E-waste in Brazil, through sectoral agreements. The authors used the Strategic Options of Development Analysis, in order to consider the conflicting viewpoints of the stakeholders and propose actions	Guarnieri et al. (2016)
Brazil	This paper studies the reverse logistics credits of E-waste, in the context of the implementation of the Brazilian Policy of Waste Management and sectoral agreement. The authors propose an analogy with carbon credits	Caiado et al. (2017).
Brazil	This paper proposes a closed cycle model to assess the impact of Brazilian public policies related to solid waste management on the social inclusion of waste pickers. The influence of these legal motivators, as well as the bargaining power of associations and cooperatives on the effective formalization of the waste pickers in the process of E-waste recovery, is studied through a dynamic simulation model	Ghisolfi et al. (2017)
India	This paper proposes a new conceptual framework of 'public understandings of E-waste and its disposal' in urban India, based on the theory of planned behaviour and conspicuous consumption. A case study carried out in the city of Bangalore to test the framework	Borthakur and Govind (2017)

(continued)

Table 5.1 (continued)

Country	Main aspects highlighted	Authors
India	This paper develops a model for forecasting product returns to the forecasting model for product returns. A case of Indian mobile manufacturing company is discussed for the validation of this model. A survey conducted by the company and findings from previous research were used for data collection on probabilities and product life cycle through Graphical Evaluation and Review Technique	Agrawal et al. (2014)
India	This paper approaches the E-waste management in India and its effects and presents a management model, for E-waste management, which considers the financial dependency of the informal sector on recycling. The model initializes an interdependent system of recycling, thus making it profitable for every concerning party involved in the recycling process	Shirodkar and Terkar (2017)

ture. Bouzon et al. (2016) researched reverse logistics development in Brazil. The authors identified 36 barriers, which were classified into seven categories: (i) technology and infrastructure, (ii) governance and supply chain process, (iii) economic, (iv) knowledge, (v) policy, (vi) market and competitors and (vii) management. Similarly, Prakash, Barua and Pandya (2015) found 28 barriers to implementation of reverse logistics in Indian electronics industry, which was categorized in (i) strategic, (ii) economic, (iii) policy, (iv) infrastructural and (vi) market related. Among the barriers cited by these authors, we can highlight lack of coordination/collaboration, customer perception about reverse logistics, lack of specific policies and incentives, lack of infrastructure and knowledge and lack of systems to monitor returns.

Besides, the study of the United Nations Environment Programme (2009) cites barriers to the transfer of technology of sustainable E-waste recycling in South Africa, Morocco, Colombia, Mexico and Brazil, which are (i) lack of specific regulation on E-waste, (ii) low technological potential and (iii) lack of investments and business models. In Brazil, the Policy of Waste Management was enacted to regulate waste management in both public and private areas. This law establishes the criteria for the reverse logistics of E-waste and other categories of hazardous waste, i.e. herbicides, lubricant oil, lamps, batteries, tyres and products' packages. Related to the technology barrier, the collecting and recycling processes are concentrated in components and materials with high added value (as circuit boards and stainless steel), but other components are undervalued. Consequently, they are discarded in improper places. The recycling of E-waste does not seem to be a high priority in the cited countries; besides that, an additional fee for the E-waste recycling sounds very unpopular.

Yacob et al. (2012) identified barriers to reverse logistics practices in Malaysian micro and small enterprises, and, according to the authors, the absence of a minimum infrastructure, such as storage locations of the returned products, transport network for collection or points of reception and a structure for product recovery, among others, compromises the implementation and execution of reverse logistics. It is important to note that establishing the adequate infrastructure to collect final consumers' products requires significant investments, which are considered as obstacles by micro and small enterprises (Yacob et al. 2012).

It is also important to emphasize that recyclers of E-waste face uncertainty in terms of the quantity and the quality. Moreover, they must adjust their decisions through some form of the dynamic process associated with the sources of the E-waste (Nagurney and Toyasaki 2005). Nonetheless, company policies and organizational structures are often obstacles, which make it difficult to change to a sustainable vision. In the case of reverse logistics, the absence of a clear view of the process further corroborates this increase. An example is the companies that have the production based on virgin raw material that consequently will not handle returned products nor recover the value of these products (Ravi and Shankar 2005).

In general, the developing countries have an available workforce in waste picking, but the infrastructure is poor and the E-waste is handled in a primitive way (Kumar et al. 2017). On the other hand, the European regulation on E-waste is a pioneer and considered the basis for other countries' policies (Restriction of Certain Hazardous Substances and Waste Electrical and Electronic Equipment directives). However, Herat and Agamuthu (2012) indicate that the transfer of appropriate technology to developing countries to manage their E-waste problem is a challenge. For the authors, it should be undertaken, keeping in mind their economic, social and environmental boundaries.

While Gomes et al. (2011) highlighted the main costs to be considered in the development of reverse logistics for E-waste, freight costs, processing costs (transportation, recycling and disposal), storage costs and non-collection costs, which reaffirm the importance of economic issues in this context, political maturity reveals the level of commitment to social and environmental issues. In environmental aspects, its degradation and resources depletion are the main reasons for sustainable choices. However, the countries' specificities still guide the main decision making. While African countries have the lowest E-waste generation ratio and European countries have the highest one, the Asian countries, except for China that presents the huge E-waste generation of the world, seem to be the intermediate points (Baldé et al. 2017).

To sum up, Abdulrahman et al. (2014) agree that the benefits of reverse logistics, which is an important tool for the circular economy, are not in practise in the emerging economies, and some steps are to be achieved yet, such as the physical structure, information technology, taxation concerns and others.

5.3 Methodological Procedures and Techniques

This chapter was elaborated based on a literature review and documental analysis related to practices on E-waste management in developing countries. It is a descriptive and exploratory research, with a qualitative approach.

The literature review considered papers published in the last 10 years in scientific databases like Science Direct, Emerald, Taylor and Francis, Elsevier and Web of

Science. Besides that, the documental research considered studies from organizations such as the United Nations Environment Programme reports from associations representing the electronics industry and websites from governments of developing countries. These documents provided information about statistics, procedures and legislation related to E-waste.

The analyses of data gathered from papers, reports, statistics and legislation were analysed through the technique of content analysis, proposed by Bardin (1977), to identify what the elements have in common to group them into categories defined a priori and/or posteriori. This technique of analysis is based on three stages: pre-analysis, exploration of the material and the treatment of the results composed of their inferences and interpretations (Bardin 1977). In this chapter, we define the categories a posteriori.

We present the data related to developing countries following the categorization: Asian countries (China, India, Russia, Pakistan, Turkey, Indonesia, Thailand and Singapore), American countries (Brazil, Chile, Argentina and Mexico) and South Africa. We do not exhaust research of the countries due to the inexistence of data and papers published related to some countries.

5.4 Presentation of Results

5.4.1 Practices Related to the Reverse Logistics of E-waste in Developing Countries

In terms of E-waste generation worldwide, in North America, the United States leads the ranking with a production of 9.4 million tonnes per year; in Asia, China leads the ranking generating 7.3 million tonnes per year; in Europe, Germany leads generating 1.9 million tonnes per year; in South America, Brazil leads generating 1.5 million tonnes per year; in Oceania, Australia leads producing 0.57 million tonnes per year; and in the African continent, South Africa leads generating 0.34 million tonnes per year (Araujo et al. 2015). Table 5.2 shows the generation of E-waste in the main developing countries.

Post-consumer electrical and electronic equipment can be reintroduced in the usage phase of the product life cycle after refurbishing or remanufacturing. In the end-of-life phase, the product and respective materials are processed to remove or decontaminate all hazardous compounds, and the safe parts can be recycled or valuable materials can be recovered (Bakhiyi et al. 2018).

The high cost of proper disposal in developed countries boosts the waste electrical and electronic-waste (E-waste) export to developing countries. The main E-waste receiving sites are Karachi (Pakistan), Guiyu (China) and Mumbai, Ahmedabad and Madras (India), whereas Taiwan acts as a redistribution destination for Japan and South Korea E-waste and Singapore for E-waste originating from North America and Europe (Imran et al. 2017). The E-waste transboundary movement leads to a scenario where the responsibilities and impacts are intertwined. According to

Table 5.2 E-waste generation in developing countries in 2016

Country	E-waste generation (kilotonnes)	E-waste regulation in force
Argentina	368	No
Brazil	1534	No
Chile	159	Yes
China	7211	Yes
India	1975	Yes
Russia	1392	Yes
South Africa	321	No
South Korea	665	Yes

Source: Baldé et al. (2017)

Ongondo et al. (2011), some specificities about E-waste management can be identified in different countries. Other characteristics are common between some countries, which allow allocating them in groups.

In developed countries, E-waste is mainly managed by private companies in association with the municipality, and the role of waste pickers is virtually null (Xavier and Carvalho 2014). In the Indian context, Ardi and Leisten (2016) explain the cause of the informal sector growth in E-waste management system. Except for Russia and Brazil, the other developing countries seem to present a pattern of behaviour regarding E-waste management that allows classification according to the respective continents.

5.4.1.1 American Countries

Brazil

Considering that the Brazilian electronics segment generates between 2% and 4% of the waste of its operations, unfortunately, estimations based on the Brazilian formal labour market indicate that no more than 1% of E-waste produced has an adequate environmental treatment (FEAM 2013).

E-waste recycling in Brazil exists countrywide, mainly in the southeast region (São Paulo State), which has several companies specialized in material fractions that have a high aggregated value (printed wiring boards, stainless steel, copper-containing components) (United Nations Environment Programme 2009). After processing the disassembling of the parts, the companies sell it to countries as Belgic and Japan, to recycle the noble metals.

In Brazil, like as in other developing countries, the waste pickers participate in reverse logistics of E-waste (Guarnieri and Cerqueira-Streit 2015; Ferri et al. 2015). They act in the collection and in the recovery of recyclable material (Franco and Lange 2011; Guarnieri et al. 2016). Another particularity in developing countries is the presence of the scrap dealers, which can be considered a middleman in the process. They act in the equipment recovery stage, disassembling the material received by technical assistance and reconditioning centres, then selling the parts to recovery and recycling companies. The recycling industries receive pieces from waste pickers and scrap dealers for the installation of new equipment (Franco and Lange 2011).

Another actor in the reverse logistics process that is typically found in developing countries are the non-governmental organizations (Guarnieri et al. 2016). In Brazil, the non-governmental organizations act also as a middleman in the process of reverse logistics, because not always the waste pickers have licences and infrastructure to deal with E-waste. These entities receive several donations from consumers and public and private companies, and then, the E-waste is sorted, refurbished and also sold to the final consumer or still donated to charitable institutions or schools by non-governmental organizations (Guarnieri et al. 2016).

The study of the United Nations Environment Programme (2009) states that E-waste does not seem to be a high priority for the industry association representing most of the information and communication technologies producing or assembling industries. In addition, an E-waste system with an additional recycling fee seems to be very unpopular, as the Brazilian tax system already puts high burdens on producers and consumers (United Nations Environment Programme 2009). This situation seems to be changing after the sanction of the Brazilian Policy of Waste Management, in 2010.

This policy introduces the principle of 'shared responsibility', involving producers, importers, retailers, government and final consumer to implement reverse logistics of E-waste, based on the 'polluter pays' principle (Brazil 2010). In Brazil, the Policy of Waste Management was a legal milestone for waste management, including E-waste and its components (Campos 2014).

This legal instrument to put in practise the principle of shared responsibility and the implementation of reverse logistics is the 'sectoral agreements' (Brazil 2010). However, the sectoral agreement of E-waste is still in negotiation; the government received proposals considering the call from the Ministry of Environment in 2013, which establishes the demand to return 17% of all electronics traded (Brazil 2013). The formalization of the law through the sectorial agreement has been postponed due to the difficulty in establishing responsibilities for reverse logistics activities (Ghisolfi et al. 2017). Consequently, their formalization in E-waste recovery process is still slow. The absence of bargaining power is a critical factor to ensure the effective inclusion of waste pickers' informal E-waste management.

Guarnieri et al. (2016) studied the structuration of the problem of Brazilian sectoral agreement of E-waste, using a structured methodology called Strategic Options and Development Analysis. The authors identified conflicting viewpoints of some stakeholders of the supply chain of electronics related to the implementation of reverse logistics and found some actions that need to be developed in terms of strategic, environmental, economic and social issues.

In addition, Caiado et al. (2017) studied the perspectives to implement the concept of reverse logistics credits in the context of E-waste management in Brazil. The reverse logistics credits were compared to carbon credits, and the authors found, according to the opinion of some stakeholders of the electronics supply chain, that Brazilian reverse logistics credits market still does not have any legal support to work on, no organization to control and audit the market and no support from the government. However, despite the difficulties, it can be considered as an alternative for companies to attend the demand of the call to a sectoral agreement.

Argentina

Although Argentina has initiated a national integrated E-waste management plan and a project on specific legislation on E-waste that should cover the ten E-waste categories according to the European directive (Ongondo et al. 2011), in 2005 and 2006, respectively, a limited number of companies operate return programmes for E-waste, such as televisions, computers and mobile phones.

In 2007, when over 20,000 tonnes of information technology waste has been generated, a third project was developed to establish guiding principles for companies working in electronic-waste management. However, these proposals have not yet received the political support needed to become effective (Boeni et al. 2008). Thus, significant amounts of E-waste still end up in municipal dumps with consumers not aware of the environmental consequences of E-waste (Protomastro 2007).

Chile

Copper is a strategic material in Chile because of its large mineral reserves. As the leading copper-producing country, Chile is responsible for 33% of world copper mining (Lagos et al. 2018) and for more than 55% of the global lithium resources, one of the raw materials of high technology components (Zhang et al. 2017). Both copper and lithium are classified as critical materials because of its economic importance in the world market.

Nevertheless, the copper ore grade is declining. To maintain the level of copper production, more ore should be mined, which results in more waste generation and water consumption. However, sustainable alternatives must be regarded to provide copper sources, such as E-waste material recovering.

It was estimated that 10,000 tonnes of computer would be generated in Chile in 2010 (Steubing et al. 2010), but the most recent report on this issue presents a generation of 159,000 tonnes of E-waste in Chile. Even considering that the first estimation was restricted to computers, the estimate of the report seems overestimated in relation to the first and vice versa. This condition determines the lack of reliability in the data presented by different databases.

According to Baldé et al. (2017), Argentina generated 368 kilotonnes of E-waste in 2016, while Chile generated 159 kilotonnes. However, in Chile were collected 0.7 kilotonnes (0.4%), and in Argentina were collected 11 kilotonnes (3%) of E-waste. An inefficient collection rate as observed in Chile can be considered one of the barriers to E-waste management.

Mexico

According to Estrada-Ayub and Kahhat (2014), from the United States to Mexico, there is a dynamic commerce of used products, in which perceived value and geographic location determine the discard rate of computers and the opportunities of wasting or commercialization of their products and materials.

Thus, at the national level, there is a challenge to overcome some barriers that are identified in the implementation phase of E-waste management plans (Cruz-Sotelo et al. 2017). Among them, there are (i) the criteria for approval by different federative entities, in relation to the emission permit, transportation and collection procedures, which industries must comply with to avoid inefficiencies and disparities that coexist in the management of such wastes; (ii) the need to strengthen the regulatory structure in order to define the roles and responsibilities of stakeholders, stakeholder groups and their responsibilities (principle of shared responsibility that is established in the waste law); (iii) modification of the E-waste disposal practice mixed with the urban waste stream; (iv) to obtain a precise estimate of E-waste, since data related to this trade are scarce, mainly due to the informality and quantity of companies involved; and (v) technological change, associated with obsolescence, in which it is necessary to create policies that respond to this diversity and can avoid unforeseen problems and stimulate solutions (Cruz-Sotelo et al. 2017; Estrada-Ayub and Kahhat 2014).

With respect to the last item, Cruz-Sotelo et al. (2017) highlight that some of the main problems are (1) lack of infrastructure, especially during the preprocessing and treatment of E-waste, (2) the absence of innovation facilities and technologies, (3) lack of investment, (4) high management costs and (5) social and security problems.

Even so, Alcántara-Concepción et al. (2016) emphasize that in Mexico, there is an effort to develop a strategic plan for the management of end-of-life computers, which involves prohibiting the use of open dumps, avoiding the disposal of end-of-life computers in landfills, economic incentives, developing recycling processes and/or assimilating new technologies for the recovery of metals, creating a national programme for collecting and transporting end-of-life computers to decommissioning and recycling centres and national collection targets with all stakeholders.

5.4.1.2 African Countries

South Africa

As can be seen in Bob et al. (2017), in South Africa the government departments and agencies have high inventories and, consequently, accumulation of electronic-waste throughout the country.

In part, this is because there is a recognition that E-waste offers threats and opportunities. In terms of the latter, E-waste, if well managed, can address the three challenges of job creation, poverty and inequality. On the other hand, there are many obstacles to managing end-of-life products safely and effectively (Widmer et al. 2005).

Thus, there is a serious problem with the increasing amount of electronic-waste, due mainly to the absence of policies that comply with international standards, as well as reliable data, safe recycling of E-waste in the formal sector and undesirable practices for management in a sustainable way (Herat and Agamuthu 2012; Widmer et al. 2005).

In this sense, according to the literature (Finlay and Liechti 2008; Widmer et al. 2005), some recommendations are suggested, investment in technology and skills, strengthening of adequate policies and legislation and E-waste management that ensures adherence by all stakeholders, as well as incentives in business and finance.

5.4.1.3 Asian Countries

Asian countries are adopting rudimentary management practices to deal with the high amounts of E-waste generated by domestic production or imported from industrialized countries (Herat and Agamuthu 2012). India, China, the Philippines, Hong Kong, Indonesia, Sri Lanka, Pakistan, Bangladesh, Malaysia, Vietnam and Nigeria are favourite destinations for E-waste. Herat and Agamuthu (2012) state that the operations in these countries are well documented evidencing that recovery practices are causing significant environmental and health impacts. They exemplify the significant amounts of E-waste containing hazardous materials can be seen dumped in open land and waterways.

According to Baldé et al. (2017), Western Asia generates 2 metric tonnes of E-waste. Constituted by high- and middle-income countries, this subregion has three countries with national E-waste legislation (Cyprus, Israel and Turkey). In this area, only 6% of E-waste is reported to be collected and recycled, mainly by Turkey. These authors indicate that other countries or private companies are supporting E-waste recycling. A facility is being built to process 39 kilotonnes of electronic-waste annually.

Firms in Malaysia should be pressured to improve reverse logistics into their operations to comply with legislation or directives introduced by foreign countries. However, this pressure was not enough to motivate manufacturers to invest in reverse logistics, due to the low level of adoption (Abdullah and Yaakub 2015).

China

China is widely known for its huge E-waste generation (Baldé et al. 2017) and E-waste management through informal treatment sites (Tang et al. 2015). Habuer et al. (2014) emphasize the negative potential impact in environment and health from E-waste management in China and the pressure for material recovering from E-waste as an alternative for natural resources shortage through urban mining techniques implementation.

According to Wang et al. (2013), China is one of the top destinations of E-waste imported from various parts of the world in the form of post-consumer products. Although this procedure is restricted by the 1989 Basel Convention, significant volumes of E-waste are processed in China. This is the case observed in the Chinese Guiyu area, where there are reports of high exposure of residents, including children, to toxic agents (Wong et al. 2007; Liu et al. 2018).

Some structural constraints in China contribute to the perpetuation of primitive E-waste processing, such as cheap workforce availability, a huge population and not strict environmental regulation (Habuer et al. 2014). Eugster and Fu (2004) state that E-waste recycling in China is supported by the informal sector and results in occupational health risks, besides being an inefficient process (Steubing et al. 2010).

Some authors consider the feasibility of a Chinese E-waste management model that allows the formal model and the informal model compatibility, prioritizing dismantling and recycling (formal) and collection (informal) (Yu et al. 2010; Chi et al. 2014).

Despite the major challenges that China still poses in the international scenario of E-waste management, one must recognize its advances in regulating the management of information on the harmful impacts generated by E-waste (Scruggs et al. 2016).

South Korea

Since 1992 it was introduced in South Korea (or the Republic of Korea) the act of saving and recycling of resources under which charges and deposit fee system were established for industries to promote recycling for different products (Rhee 2016).

In 2003 it was instituted a mandatory E-waste recycling programme in South Korea that included 11 categories of post-consumer products (Kim et al. 2013), but since 2013 there are only five groups in which there are 27 categories of E-waste (Rhee 2016). This programme was based on the extended producer responsibility principle, and since 2006 the brominated flame retardants are restricted in several products from South Korea. These same authors estimated the E-waste generation amounts according to eight categories and suggested that by 2020 there will be more than 70 million units of mobile phones ownership, with an average lifespan of 2.5 years, and an expected E-waste generation of 20 million units in 2020.

According to Park et al. (2014), Korea has seven recycling centres where E-waste materials are processed. These authors analysed emissions of flame retardants compounds, such as polybrominated diphenyl ethers, during usage, disposal and recycling life cycle phases in Korea. Despite being banned in the Fourth Conference of the Parties in Stockholm, these compounds are still present in different E-waste components. Park et al. (2014) identified significant amounts of polybrominated diphenyl ether emissions from plastic recycling plants. The findings suggest that old televisions devices, before the year of 2000, have ten times more polybrominated diphenyl ethers than those produced after this year, and the storage in opening areas also suggested higher impact from emissions.

A dilemma has been established as to the definition of the percentage to be stipulated for the management of E-waste. While a proposal stipulated the percentage according to the amount of electrical and electronic equipment placed on the market annually, another proposal seeks to establish the percentage according to the volume of E-waste generated annually (Kim et al. 2013).

E-waste generation increased slightly in Korea from 2003 to 2007, but in 2009 it stagnated due to the economic depression (Rhee 2016). This author presents that Korea collected in 2013 almost 160 million tonnes of E-waste, from which over 60% were collected by producers in reverse logistics systems and more than 25% were collected by recycling companies.

It was developed in Korea in 2001 the Allbaro System that accomplishes the waste monitoring, including the transboundary movement (import and export) of hazardous waste. This can be considered an innovative initiative that provides support for E-waste management.

India

In urban India, one of the major concerns has been the responsible management of E-waste, considering the damages to human health and the environment (Borthakur and Govind 2017). The growth of population in urban India from 11.4% in 1901 to 31.16% in 2011 resulted in an alarming increase in waste generation (Yadav et al. 2016). Besides that, India, along with Africa and China, has been the main destinations of E-waste generated in developed countries (United Nations Environment Programme 2009; Agoramoorthy and Chakraborty 2012).

Due to the illegal imports, it is observed an increase of domestic generation of E-waste, and it is estimated that 90% of all the million tonnes of E-waste of India are discarded by informal recyclers who use rudimentary methods to process E-waste (Shirodkar and Terkar 2017).

Agrawal et al. (2014) point out that the Confederation of Indian Industries estimates that the electronics industry in India has a market size of USD 65 billion and is expected to reach USD 400 billion by the year 2020.

Indian industry is dependent upon import electronics hardware, with commercial transaction around USD 18.61 billion in 2010–2011. For this reason, the manufacturers do not have technology and structure to recycling or reconditioning many components. Besides that, the cost to send them back to suppliers is unfeasible (Agrawal et al. 2014).

The Confederation of Indian Industries estimates in 2011 that the total E-waste generated in India is around 146,000 tonnes per year, and it is growing at the average of 10% every year (Agrawal et al. 2014). At the same time, there are very little or no regulations or laws regarding E-waste management in India (Shirodkar and Terkar 2017).

Agoramoorthy and Chakraborty (2012) presented a ranking of E-waste-generating states in India, in which 65 cities are responsible for the 70% of the amount of production of E-waste. The Maharashtra state led the ranking, followed by Tamil Nadu, Andhra Pradesh, Uttar Pradesh, West Bengal, Delhi, Karnataka, Gujarat, Madhya Pradesh and Punjab. Among the cities, Mumbai ranks first followed by Delhi and Bangalore.

Russia

With a population of more than 140 million people, Russia annually produces about 1.4 million tonnes of E-waste and has a per capita generation of 9.7 kilos in 2016 (Baldé et al. 2017). Since January 2015 entered into force a regulation that establishes the obligation of producers and importers to take back some products and packages establishing (i) their own waste management infrastructure, (ii) collective organizations to manage waste, (iii) contracts with regional operators or (iv) paying the environmental fee according to average cost of collection, transportation and disposal of each product.

Among the 36 categories listed in the federal law amendment, 10 are identified as E-waste, as detailed below:

- (24) Computers and peripherals
- (25) Communications equipment
- (26) Consumer electronics
- (27) Optical devices and photographic equipment
- (30) Electric lighting equipment
- (31) Household electric devices
- (32) Non-electric household devices
- (33) Power-assisted hand tools
- (34) Industrial refrigeration and ventilation equipment
- (35) General purpose machinery and equipment, not included in other categories

The Group of Eight is a group made up of Canada, France, Germany, Italy, Japan, Russia (suspended), the United Kingdom and the United States. The Group of Eight 3Rs Initiative (Reduce, Reuse and Recycle) was introduced by Japan during the Group of Eight Summit in June 2004. During the Asia 3Rs Conference held in Tokyo during November 2006 where 20 Asian countries, 6 Group of Eight countries and 8 international organizations participated, progress and issues related to environmentally sound management of E-waste in the Asian region were discussed, and delegates from Asian countries and experts made presentations on case studies of E-waste management (United Nations Centre for Regional Development 2011).

Turkey

With 78.97 million inhabitants, Turkey generates about 623 thousand tonnes of E-waste in 2016, which correspond to 7.9 kilos per capita (Baldé et al. 2017). It was expected an increasing average growth rate of E-waste generated about 5% per year by the Regulatory Impact Analysis report prepared in 2011 (Regional Environmental Center of Turkey 2011). However, the E-waste per capita estimated in 2020 in this study is already surpassed according to Baldé et al. (2017). This amount is also far from the target of 20 kilo/capita to be achieved in 2019 (Camgöz-Akdag and Aksoy 2014).

According to Gök et al. (2017), there are few companies in the E-waste reverse logistics in Turkey. Two firms collect and handle E-waste. It is collected in special containers and licenced transporters placed in workplaces by these firms. Furthermore, collected E-waste is separated into plastics and metal parts and forwarded to recycling processes. These companies are not capable of recycling all accumulated E-waste. Fluorescence toner cartages and capacitors are recycled in other countries. The waste that cannot be recycled is sent to disposal facilities (Gök et al. 2017).

Turkey implemented the extended producer responsibility for E-waste in 2012, the so-called Regulation on the Control of Waste Electrical and Electronic Equipment number 28300. With this regulation, most of the E-waste arising is collected by different operators which minimized the incorrect disposal in municipal waste dumps. Municipalities were preparing an E-waste management plan in 2014 including public information about E-waste collection programme and E-waste collection. The Ministry of Environment and Urban Planning should evaluate the plan conformity (Camgöz-Akdag and Aksoy 2014).

As Turkey is not yet a European Member State, there is no transposition of European legislation in place. The great increase of the E-waste volume in Turkey and the desire to participate in Europe impulse the regulation in this country. Kilic et al. (2015) indicate the following problems to implement the directive: (i) Turkey has less E-waste amount compared to European countries; (ii) the amount of E-waste varies across the country; (iii) Turkey does not have separate collection infrastructure; (iv) Turkey does not possess proper treatment facilities for cooling and freezing equipment; (v) the E-waste recycling sector is highly dominated by informal scrap dealers; (vi) Turkey does not still have the awareness for proper collection and treatment of E-waste; and (vii) there is inadequate intellectual knowledge for technical and financial sides. Camgöz-Akdag and Aksoy (2014) also indicate the challenges faced by Turkey to improve their E-waste reverse logistics system and propose some recommendations for improving it.

Indonesia

Due to the high economic growth as well as the rapid development of technology, E-waste generated in Indonesia is expected to increase significantly (Andarani and Goto (2014). Thus, there are indications that a large reuse flow is currently occurring and effective inventory and management of E-waste remain a challenge for Indonesia (Rochman et al. 2017).

In addition to these challenges, the Indonesian government faces two major problems: first, E-waste is smuggled into the country in the form of user devices; second, a large number of second-hand devices with unknown sources circulate in the country without control (Panambunan-Ferse and Breiter 2013). In part this may be since Indonesia is made up of thousands of islands and consequent difficulty in monitoring and control of ports and boats, in addition to the long chain in the primary and secondary markets, as well as informal and formal channels for the management of this waste (Rochman et al. 2017), making the mapping of E-waste streams quite complex. Therefore, Indonesia is suspected of receiving a large share of this waste through illegal imports (Anderson 2010).

Thus, as can be seen in the literature (Panambunan-Ferse and Breiter 2013; Rochman et al. 2017), the problems of E-waste management in Indonesia are due to lack of infrastructure and a system of information to quantify, monitor and handle electronic-waste and the absence of strong and effective legal regulations. As actions to mitigate these problems, we have adopted an incentive system for cellular recovery, the inclusion of informal actors in the development of regulations and adequate management of electronic-waste, as they represent a significant component of the current waste recycling system and whose profit margins are low and exposed to risk.

Thailand

Thailand has 68.98 million inhabitants in 2016. In this year, the country generates about 507 thousand tonnes of E-waste, which correspond to 7.4 kilos per capita

(Baldé et al. 2017). Chirapat et al. (2012) point out several problems associated with E-waste, such as inefficient domestic collecting and dismantling systems, low-quality products, inadequate E-waste management regulations and limited expertise and technology. In this same sense, Herat and Agamuthu (2012) and Baldé et al. (2017) add the lack of general awareness about E-waste and incomplete databases and inventories related to this waste as other barriers in the country. However, as a great driver for reverse logistics, the absence of a national regulation is a significant challenge.

Thai government passed the National Strategic Plan on Integrated Management of E-waste (E-waste Strategic Plan) in July 2007, but this plan has implementation problems. According to Chirapat et al. (2012), there is a guideline approved by the government to implement an E-waste management plan for the entire country. This guideline includes application of precautionary principle and polluter pay principle to producers and importers responsible for E-waste management; implementation of regulations to reduce, reuse and recycle domestic and imported products; establishment of economic, financing and marketing mechanisms that motivate and stimulate E-waste management; research and development of eco-friendly production technology; capacity building of local administrative authorities; and other actions.

The Thai government proposed a Green Industry Project in 2014. Kamolkittiwong and Phruksaphanrat (2015) investigated the main critical drivers for the successful adoption of companies in the country. Regulation is the most important driver among the entire identified green supply chain management drivers studied. This result reinforces the importance of the regulation enforcement to E-waste management improvement.

Singapore

With 5.59 million inhabitants, Singapore generates about 100 thousand tonnes of E-waste in 2016, which correspond to 17.9 kilos per capita (Baldé et al. 2017). This country is indicated as a key node in the global networks of E-waste trade and traffic. The Basel Convention on the Transboundary Shipment of Hazardous Waste segregates countries into two groups. One group, Annex VII parties (comprising developed countries), is accused of exporting hazardous waste to the non-Annex VII parties (Lepawsky and Connolly 2016). Singapore, a non-Annex VII party, is a major transhipment hub for waste electronics discarded to other countries in Asia (Connolly 2012). Lepawsky and Connolly (2016) highlight that, among other facilities, foreign traders are attracted to Singapore because of the high quality of electronics discarded there. The authors also emphasize as a non-Annex VII party, there is nothing in the Basel Convention prohibiting the country to export E-waste to other non-Annex VII countries.

Pakistan

According to Iqbal et al. (2015) and Baldé et al. (2017), there is no reliable data available on the E-waste generation, imported, recycled or dumped. It is estimated that the country generates about 301 thousand tonnes of E-waste, which correspond to 1.6 kilos per capita (Baldé et al. 2017). Pakistan receives thousands of tonnes of

E-waste from developed countries like the United States and Europe (redistributed by Singapore or Taiwan) because it is a very cheap site for E-waste recycling. Karachi receives the major portion of E-waste imported to Pakistan (89.39% in total estimated) followed by Lahore (Imran et al. 2017).

To avoid the entrance of E-waste in Pakistan, the government prohibited E-waste imports. It is also a signatory to the Basel Convention and has its own relevant legislation regarding E-waste imports: the Pakistan Environmental Protection Act (1997), Pakistan National Environment Policy (2005) and Pakistan Import Policy Order (2016). However, they are still being imported as second-hand items or illegally. The illegal import is also estimated in an annual average of 95.4 kilotonnes, constituted mostly by computers and related products (Imran et al. 2017).

There is no registered recycling facility in the country. Precarious recycling practices like physical dismantling, open burning, acid bathing and use of blowtorches are performed in open air and in small workshops. Hundreds of workers including teenage children work in dismantling and extraction of valuable materials. Consequently, damaging practices have been used for recycling E-waste, but governmental and non-governmental organizations have not been dedicated to solving this problem (Iqbal et al. 2015). Imran et al. (2017) also emphasize that the recycling work is manually realized without protective equipment.

To improve this scenario, Iqbal et al. (2015) suggest the enforcement of the existent laws, such as the Basel Convention, as a first action followed by the enactment of specific national E-waste legislation. However, Imran et al. (2017) state that the existing legislation is not sufficient for the challenge imposed. They suggest the necessity to establish controls in imports, environmental management (handling, storing, sorting and transporting of discarded items), controls for illegal imports and adequate E-waste workers health conditions to better support manufacturers, employees and other stakeholders.

5.4.2 Results Discussion

After analysing the main characteristics of E-waste management in developing countries, it was possible to categorize some barriers that avoid the efficiency in this process, cited by the articles analysed, which are presented in Table 5.3.

In our study, the most prominent barriers are those related to management and political issues. Some other researches related to barriers to adoption of reverse logistics were published in the last years.

Baldé et al. (2017) highlight the lack of reliable E-waste data at the country level, which we can verify in this research. The Global E-waste Monitor 2017 found that only 41 countries in the world collect international statistics on E-waste.

Abdulrahman et al. (2014) studied the barriers of reverse logistics in China and found four categories: (i) management; (ii) financial; (iii) policy and (iv) infrastructure. In addition, Prakash and Barua (2015) performed a study of barriers in India, related to E-waste reverse logistics. The authors found barriers similar to what we

Table 5.3 Categorization of main barriers to the adoption of reverse logistics in developing countries

Description of barriers	Author (year)	Country
Financial and economic barriers		
Limited funding has also caused significant impediments to the effective management of toxic wastes	Andarani and Goto (2014)	Indonesia
Lack of financial resources	Aydin Temel et al. (2018)	Turkey
Legal barriers		
The legislation is required to establish: Controls in imports, environmental management (handling, storing, sorting and transporting of discarded items), controls for illegal imports, adequate E-waste workers health conditions	Imran et al. (2017)	Pakistan
Market-related barriers		
The amount of used electrical and electronic equipment's imported into the country has increased with varying consequences for businesses and the local economy	Amankwah-Amoah (2016)	Mexico
The lack of definition related to the response to support the implementation of E-waste reverse logistics network	Guarnieri et al. (2016) and Caiado et al. (2017)	Brazil
Environmental barriers		
Primitive E-waste processing resulting in environmental and health impacts	Wong et al. (2007) Liu et al. (2018)	China
Toxic agent's exposure to E-waste handling	Pascale et al. (2016)	South America
Primitive E-waste management	Habuer et al. (2014)	China
In Indonesia, it is generally difficult to find any E-waste dumped in official final disposal sites or landfill	Andarani and Goto (2014)	Indonesia
Significant amounts of E-waste still end up in municipal dumps with consumers not keenly aware of the environmental consequences of E-waste	Protomastro (2009)	Argentina
Policy-related barriers		
Regulation is the most important challenge to improving the E-waste reverse logistics in Thailand. Even if there is some laws or projects, they need to be enforced to boost E-waste management	Kamolkittiwong and Phruksaphanrat (2015) Chirapat et al. (2012)	Thailand
There is a lack of interest from the government to control the E-waste issue and to invest in research in this field	Imran et al. (2017)	Pakistan
Corruption and ineffective data collection and dissemination on the material flow of electrical and electronic equipment and E-waste are also hurdled to overcome in the developing countries	Osibanjo and Nnorom (2007)	Africa
There is no specific legislation that deals with E-waste in South Africa	Lombard (2004)	South Africa

(continued)

Table 5.3 (continued)

Description of barriers	Author (year)	Country
There is no specific regulation regarding E-waste; hence, national regulations have not identified E-waste terminology	Andarani and Goto (2014)	Indonesia
The lack of effectiveness in monitoring the compliance with the Brazilian solid waste management	Guarnieri et al. (2016) and Caiado et al. (2017)	Brazil
Insufficient policies and deficiency of political priorities, poor coordination between authorities	Aydin Temel et al. (2018)	Turkey
Management-related barriers		
The cheap workforce in E-waste management	Habuer et al. (2014)	China
Lack of data availability on E-waste management	Ongondo et al. (2011), Kim et al. (2013) and Steubing et al. (2010)	Developing countries Korea
Lack of data reliability	Steubing et al. (2010)	Developing countries
There are a very few who hold expertise in the field of E-waste management	Imran et al. (2017)	Pakistan
Semi-formal recycling of E-waste; the dearth of data on the generation of E-waste	Osibanjo and Nnorom (2007)	South Africa
Formal recycling of E-waste using efficient technologies and facilities is rare	Andarani and Goto (2014)	Indonesia
Limited infrastructure for the flow of recycled computers and obtaining recycled materials	Alcántara-Concepción et al. (2016)	Mexico
There is no assignment of the functions of each part and no regulation on the creation of a fund for these activities or the parties responsible for the economic contributions to make tasks for the management of end-of-life computers and other E-waste	Alcántara-Concepción et al. (2016)	Mexico
Significant prolongation of the life cycle of the equipment. Insufficient formal mechanisms for handling E-waste. The absence of management of electronic-waste in rural areas. The absence of extended responsibility for the producer is another problem	Cruz-Sotelo et al. (2016)	Mexico
The current capacity of E-waste recyclers is not considered high enough to absorb the potential E-waste quantities should decommissioned technology that is being stored be released into the waste stream	Finlay (2005)	South Africa
Knowledge-related barriers		
There is a lack of first-hand knowledge regarding the amount of E-waste imported to the country. The Pakistan regulation bans all the hazardous items restricted by the Basel Convention but does not specify the terms 'E-waste, electronic-waste and electric and electronics equipment waste'. The data on imported good does not mention this	Imran et al. (2017)	Pakistan

(continued)

Table 5.3 (continued)

Description of barriers	Author (year)	Country
A lack of general awareness of E-waste among the public	Finlay (2005)	South Africa
Lack of information on production, management and recycling of E-waste	Baldé et al. (2017)	Global
Technical and technological barriers		
Lack of technology for the recycling of high added value components such as printed circuit boards, which contain precious metals, rare earth metals and heavy metals. The feasibility of recycling these materials today is dependent on the high logistics cost that is added to the cost of recycling in other countries	Ghisolfi et al. (2017)	Brazil
Weak technical resources and poor infrastructure	Aydin Temel et al. (2018)	Turkey

found in this research. They categorize barriers as legal, organizational, economic, management, technological, infrastructural and market-related barriers.

Bouzon et al. (2016), studying the context of Brazil, identified 36 barriers to the implementation of reverse logistics, which were categorized as (i) technology- and infrastructure-related issues, (ii) governance- and supply chain process-related issues, (iii) economic-related issues, (iv) knowledge-related issues, (v) policy-related issues, (vi) market- and competitor-related issues and (vii) management-related issues. It is important to emphasize that the authors also used a multicriteria decision aid method to prioritize the barriers.

On the other hand, Prakash and Barua (2015) applied a multicriteria decision aid method, in order to find alternatives that can overcome these barriers. They found 20 solutions gathered from relevant literature and expert views, which can be viewed in Table 5.4.

Guarnieri et al. (2016) studied the problem of implementation of E-waste reverse logistics in Brazil, through a problem structuring method, called Strategic Options and Development Analysis. Similarly to Bouzon et al. (2016), these authors found some actions that can be implemented in order to overcome some barriers, which are presented in Fig. 5.1

So, some developing countries are being able to propose measures to overcome these barriers. One of them could be pointed out; the Global E-waste Monitor 2017 found that, although the E-waste challenge is on the rise, a growing number of countries are adopting E-waste legislation. Baldé et al. (2017) state that 66% of the world population is covered by national E-waste management laws, which means an increase from 44% that were covered in 2014. The authors state that the main challenge with sustainable E-waste management in Latin America is the acceleration of all legislation processes. It is important to highlight that, considering the countries approached in this research, only Brazil in Latin America has a legislation, which is the Brazilian Policy of Solid Waste (Brazil 2010, Law. 12,305/2010).

Table 5.4 Solutions to overcome barriers to adopt reverse logistics

Solutions to overcome barriers to adopt reverse logistics			
i)	Top management awareness and support	xii)	Control over turnaround times
ii)	Balancing cost efficiency with customer responsiveness	xiii)	Create public awareness of environmental issues and conservation
iii)	Simplified and standardized processes	xiv)	Enforce environmental legislation, regulations and directives
iv)	The detailed insight into cost and performance	xv)	Develop infrastructure support and facility
v)	Cross-functional collaboration	xvi)	Implement green practices for electronic products
vi)	Strategic collaboration with reverse chain partners	xvii)	Create, develop and invest in reverse logistics technology
vii)	Aligned policies and processes		
viii)	Strategic focus on avoiding returns	xviii)	Make e-collaboration for fast and effective coordination among supply chain members
ix)	Perceive returns as perishable goods	xix)	Develop a closed-loop supply chain by integrating reverse logistics
x)	Reverse logistics as part of the sustainability programme	xx)	Develop outsourcing strategy for recovery and collection of end-of-life products
xi)	Reclaiming value from returns		

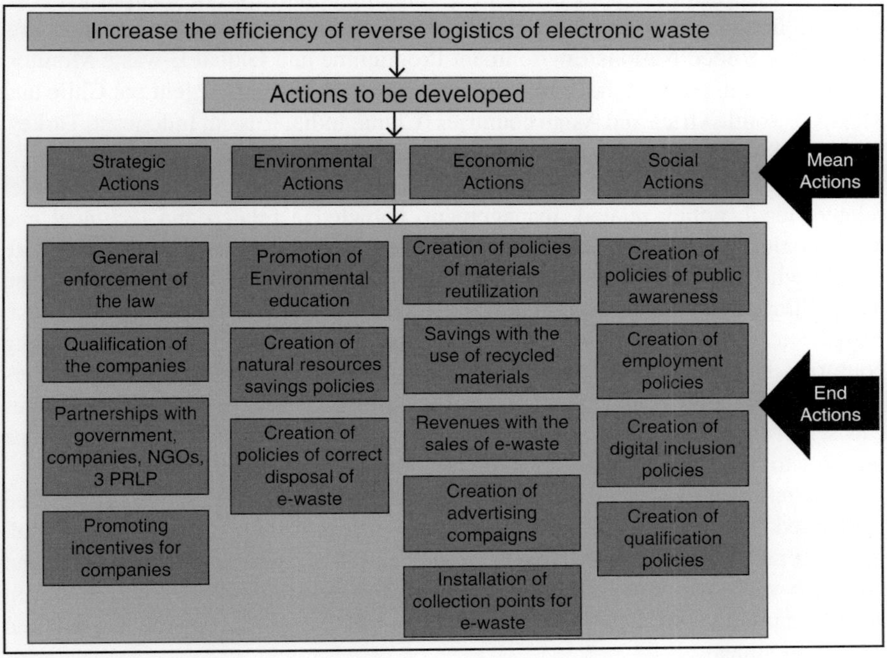

Fig. 5.1 Actions to be developed to overcome barriers to the adoption of E-waste reverse logistics. (Source: Guarnieri et al. (2016))

The major receiving E-waste countries of the world collect obsolete electronic goods mainly that are sent to the recycling industry. Only 2% of them are reused (Imran et al. 2017). To avoid these problems, Sri Lanka enacted a policy to encourage the import of high-quality products and materials to minimize the E-waste, and they have internal facilities for the collection of discarded items. China is considered to have a limited legal framework in this regard and has banned the import of dozens of electronic items since 2000 (Imran et al. 2017).

For instance, China, the leader in E-waste production, began to stand out in minimizing barriers to change E-waste import and conventional recycling activities, also known as backyard recycling. This country also employs extended producer responsibility that encourages manufacturers to take back, recover and dispose of the waste (Chung et al. 2011). However, a lot of work is still necessary to overcome all the challenges indicated. The authors agree that the enforcement of regulatory issues is the most important action to achieve this goal in developing countries.

5.5 Final Considerations

This chapter discussed the recycling challenges regarding the adoption of E-waste reverse logistics, related to the main barriers and possible solutions to overcome these barriers, from the perspective of developing countries. For this purpose, we gathered information from papers published in international databases and reports such as the United Nations Environment Programme and Global E-waste Monitor, thus identifying data available to American countries (Brazil, Argentina, Chile and Mexico), South Africa and Asian countries (China, India, Russia, Indonesia, Turkey, Pakistan, South Korea, Thailand and Singapore).

We could categorize the barriers into financial/economics, environmental, market related, legal, policy related, management, knowledge related and technical and technological related. Our categorization was performed similarly by Bouzon et al. (2016), who studied the Brazilian context; Prakash and Barua (2015), who performed a modelling study; and Abdulrahman et al. (2014), who studied the Chinese context. Nevertheless, our approach was conducted based on the reality of several developing countries. It is important to point out that in our study, the most prominent barriers are those related to management and political issues. These barriers or challenges can serve as input to find actions or alternatives to overcome them. It was also possible to find some opportunities for solutions that can be implemented.

It is important to emphasize that this research differentiates from those performed by United Nations Environment Programme report (2009), Abdulrahman et al. (2014), Prakash and Barua (2015), Bouzon et al. (2016), Guarnieri et al. (2016) and Global E-waste Monitor (2017) due to focuses in developing countries and not in specific countries or based on a global perspective (including developed and developing ones).

We also find that the reverse logistics concept, that can be understood as the management of returns, including the sorting of residues, collection, storage, warehousing and delivering processes, besides the information associated makes possible the recycling, reconditioning and remanufacturing activities. The reverse logistics is also recognized as an important part of the circular economy, providing the revaluation of the electronic residues, inserting them again in a new production process. The circular economy is based on the idea of using the waste as a raw material in an infinite cycle. In addition, the concept of urban mining should be highlighted due to the involved process of the extraction of precious metals and valuable parts of electronics to recycle them, instead of extracting them from the environment. The three concepts approached in this chapter complement each other to provide a full solution, considering since the design of the product until the reinsertion of the residue generated in the productive cycle again.

The main limitations of this research are related to the unavailability of data related to E-waste management in developing countries, an issue already found in the Global E-waste Monitor (2017) and United Nations Environment Programme report (2009). We also must emphasize that this chapter was based on a literature review and documental analysis and does not contain any empirical result.

However, these limitations can conduct opportunities for further research related to E-waste reverse logistics and management. Firstly, empirical studies at national levels in developing countries can be carried out. Secondly, studies based on empirical results should be performed to compare the barriers and solutions in developing countries. Thirdly, studies approaching the best practices of E-waste reverse logistics related to developed countries, in terms of management, technological, legal, infrastructure, financial and economic, knowledge and market, should be used as a basis for the proposition of models to overcome barriers in developing countries. Fourthly, studies using problem structuring and multicriteria methods can be performed to structure and prioritize the barriers and actions to be developed in order to overcome the barriers in developing countries.

The contributions of this chapter are twofold. First, we gathered relevant information related to recycling challenges of E-waste in developing countries, and we categorize them in types of barriers. Second, the identification of some solutions and actions to overcome these barriers is also performed, which can be useful for practitioners and researchers. The managers can use this information for better decision making, the government can develop some public policies related to barriers and solutions pointed out, and the researchers can deepen the study related to the gaps emphasized in this chapter.

Acknowledgements The authors are grateful to the National Council for Scientific and Technological Development for their support (process number: 406263/2016-7).

References

Abdullah NAHN, Yaakub S (2015) The pressure for reverse logistics adoption among manufacturers in Malaysia. Asian J Bus Account 8(1):151–178

Abdulrahman D, Gunasekaran A, Subramanian N (2014) Critical barriers in implementing reverse logistics in the Chinese manufacturing sectors. Int J Prod Econ 147(Part B):460–471. https://doi.org/10.1016/j.ijpe.2012.08.003

Achillas CH, Vlachokostas CH, Aidonis D, Moussiopoulos N, Iakovou E, Banias G (2010) Optimising reverse logistics network to support policy-making in the case of electrical and electronic equipment. Waste Manag 30:2592–2600

Agoramoorthy G, Chakraborty C (2012) Environment: control electronic waste in India. Nature 485(7398):309

Agrawal S, Singh RK, Murtaza Q (2014) Forecasting product returns for recycling in Indian electronics industry. J Adv Manag Res 11(1):102–114. https://doi.org/10.1108/JAMR-02-2013-0013

Alcántara-Concepción V, Gavilán-García A, Gavilán-García IC (2016) Environmental impacts at the end of life of computers and their management alternatives in México. J Clean Prod 131:615–628. https://doi.org/10.1016/j.jclepro.2016.04.125

Amankwah-Amoah J (2016) Global business and emerging economies: towards a new perspective on the effects of e-waste. Technol Forecast Soc Chang 105:20–26. https://doi.org/10.1016/j.techfore.2016.01.026

Andarani P, Goto N (2014) Potential e-waste generated from households in Indonesia using material flow analysis. J Mater Cycles Waste Manag 16(2):306–320. https://doi.org/10.1007/s10163-013-0191-0

Anderson M (2010) What an E-waste. IEEE Spectr 47(9):72

Appelbaum A (2002) Europe cracks down on e-waste. IEEE Spectrum (May), pp 46–51. https://doi.org/10.1109/6.999794

Araujo MVF, Oliveira UR, Marins FAS, Muniz J Jr (2015) Cost assessment and benefits of using RFID in reverse logistics of waste electrical & electronic equipment (E-WASTE). Proc Comp Sci 55:688–697. https://doi.org/10.1016/j.procs.2015.07.075

Ardi R, Leisten R (2016) Assessing the role of informal sector in E-WASTE management systems: a system dynamics approach. Waste Manag 57:3–16. https://doi.org/10.1016/j.wasman.2015.11.038

Awasthi AK, Cucchiella F, D'Adamo I, Li J, Rosa P, Terzi S, Zeng X (2018) Modelling the correlations of e-waste quantity with economic increase. Sci Total Environ 613:46–53. https://doi.org/10.1016/j.scitotenv.2017.08.288

Aydin Temel F, Konuk N, Turan NG, Ayeri T, Ardali Y (2018) The SWOT analysis for sustainable MSWM and minimization practices in Turkey. Global NEST J 20(1):83–87

Bakhiyi B, Gravel S, Ceballos D, Flynn MA, Zayed J (2018) Has the question of e-waste opened a Pandora's box? An overview of unpredictable issues and challenges. Environ Int 110:173–192. https://doi.org/10.1016/j.envint.2017.10.021

Baldé CP, Forti V, Gray V, Kuehr R, Stegmann P (2017) The global E-waste monitor – 2017, United Nations University (UNU), International Telecommunication Union (ITU) & International Solid Waste Association (ISWA), Bonn/Geneva/Vienna. Available at: https://www.itu.int/en/ITU-D/Climate-Change/Pages/Global-E-waste-Monitor-2017.aspx

Bardin L (1977) Análise de conteúdo. Lisboa: Edições 70

Bob U, Padayachee A, Gordon M, Moutlana I (2017) Enhancing innovation and technological capabilities in the management of E-waste: case study of South African government sector. Sci Technol Soc 22(2):332–349. https://doi.org/10.1177/0971721817702293

Boeni H, Silva U, Ott D (2008) E-waste recycling in Latin America: overview, challenges and potential. Focus 11(June 2014):1–10. Retrieved from http://ewasteguide.info/files/2008_Keynote_Boeni_REWAS.pdf

Borthakur A, Govind M (2017) Public understandings of E-waste and its disposal in urban India: from a review towards a conceptual framework. J Clean Prod 172:1053–1066. https://doi.org/10.1016/j.jclepro.2017.10.218

Bouzon M, Govindan K, Rodriguez CMT (2015) Reducing the extraction of minerals: reverse logistics in the machinery manufacturing industry sector in Brazil using ISM approach. Resour Policy 46:27–36

Bouzon M, Govindan K, Rodriguez CMT, Campos LM (2016) Identification and analysis of reverse logistics barriers using fuzzy Delphi method and AHP. Resour Conserv Recycl 108:182–197. https://doi.org/10.1016/j.resconrec.2015.05.021

Brazil (2010) Law 12,305, of 2 august 2010. Institutes the National Policy on Solid Waste; alters Law No. 9,605 of 12 February 1998; and makes other provisions. The Official Gazette, Brasília, Brazil

Brazil (2013) Edital 01/2013 de chamamento de acordos setoriais para a logística reversa de resíduos de equipamentos eletroeletrônicos (Proclamation 01/2013 calling for the development of sectorial agreement for the implementation of reverse logistics of consumer electronics products and their components). Available at: http://www.abras.com.br/pdf/editaleletroeletronicos.pdf

Caiado N, Guarnieri P, Xavier LH, Chaves GDLD (2017) A characterization of the Brazilian market of reverse logistic credits (RLC) and an analogy with the existing carbon credit market. Resour Conserv Recycl 118:47–59. https://doi.org/10.1016/j.resconrec.2016.11.021

Camgöz-Akdag H, Aksoy HM (2014, January) Green supply chain management for electric and electronic equipment: case study for Turkey. In: Proceedings of the international annual conference of the American Society for Engineering Management. American Society for Engineering Management (ASEM), p 1

Campos HKT (2014) Recycling in Brazil: challenges and prospects. Resour Conserv Recycl 85:130–138. https://doi.org/10.1016/j.resconrec.2013.10.017

Chi X, Wang MYL, Reuter MA (2014) E-waste collection channels and household recycling behaviors in Taizhou of China. J Clean Prod 80:87–95. https://doi.org/10.1016/j.jclepro.2014.05.056

Chirapat P, Kittinan A, Kiattiporn W, Nawanuch T, Surus T (2012, September, 1–5) Development of forecasting model for strategically planning on E-waste management in Thailand. In: Electronics Goes Green 2012+(EGG), 2012. IEEE

Chung SS, Lau KY, Zhang C (2011) Generation of and control measures for, e-waste in Hong Kong. Waste Manag 31(3):544–554. https://doi.org/10.1016/j.wasman.2010.10.003

Connolly CP (2012) Singapore is a gold mine: re-orienting international flows of secondhand electronics (Doctoral dissertation, Memorial University of Newfoundland). Department of Geography, Available at: http://research.library.mun.ca/2327/. Accessed Feb 2018

Cruz-Sotelo S, Ojeda-Benítez S, Velázquez-Victorica K, Santillán-Soto N, García-Cueto O, Taboada-Gonzalez P, Aguilar-Virgen Q (2016) Electronic waste in Mexico – challenges for sustainable management. Chapter from the book E-Waste in transition – from pollution to resource. https://doi.org/10.5772/64449

Cruz-Sotelo S, Ojeda-Benítez S, Jáuregui Sesma J, Velázquez-Victorica K, Santillán-Soto N, García-Cueto O, Alcántara C (2017) E-waste supply chain in Mexico: challenges and opportunities for sustainable management. Sustainability 9(4):503. https://doi.org/10.3390/su9040503

de Oliveira CR, Bernardes AM, Gerbase AE (2012) Collection and recycling of electronic scrap: a worldwide overview and comparison with the Brazilian situation. Waste Manag 32(8):1592–1610. https://doi.org/10.1016/j.wasman.2012.04.003

Estrada-Ayub JA, Kahhat R (2014) Decision factors for e-waste in Northern Mexico: to waste or trade. Resour Conserv Recycl 86:93–106. https://doi.org/10.1016/j.resconrec.2014.02.012

Eugster M, Fu H (2004) e-Waste assessment in PR China–A case study in Beijing. Swiss e-Waste programme. Empa-Materials Science and Technology, St. Gallen, Beijing

FEAM - Fundação Estadual do Meio Ambiente (Foundation of Environment of Minas Gerais). Diagnóstico da Geração de Resíduos Eletroeletrônicos no Estado de Minas Gerais (Diagnosis of Electrical and Electronic Waste Generation in the Minas Geraus State) (2013) Available at: http://ewasteguide.info/files/Rocha_2009_pt.pdf. Accessed 15 Nov 2017

Ferri GL, Chaves GLD, Ribeiro GM (2015) Reverse logistics network for municipal solid waste management: the inclusion of waste pickers as a Brazilian legal requirement. Waste Manag 40:173–191. https://doi.org/10.1016/j.wasman.2015.02.036

Finlay A (2005, September) E-waste challenges in developing countries: South Africa Case Study. APC Issue Papers. Association for Progressive Communications

Finlay A, Liechti D (2008) E-waste assessment South Africa. eWASA, Johannesburg

Franco RGF, Lange, LC (2011) Estimativa do fluxo dos resíduos de equipamentos elétricos e eletrônicos no município de Belo Horizonte, Minas Gerais, Brazil. (Estimation of the flow of waste electrical and electronic equipment in the city of Belo Horizonte, Minas Gerais, Brazil). Engenharia Sanitária Ambiental 16(1):73–82

Ghisolfi V, Chaves GDLD, Siman RR, Xavier LH (2017) System dynamics applied to closed loop supply chains of desktops and laptops in Brazil: a perspective for social inclusion of waste pickers. Waste Manag 60:14–31. https://doi.org/10.1016/j.wasman.2016.12.018

Gök G, Tulun Ş, Gürbüz OA (2017) Consumer behavior and policy about E-waste in Aksaray and Niğde Cities, Turkey. Clean Soil Air Water 45:1500733. https://doi.org/10.1002/clen.201500733

Gomes MI, Barbosa-Povoa AP, Novais AQ (2011) Modelling a recovery network for E-WASTE: a case study in Portugal. Waste Manag 31(7):1645–1660. https://doi.org/10.1016/j.wasman.2011.02.023

Govindan K, Soleimani H (2017) A review of reverse logistics and closed-loop supply chains: a journal of cleaner production focus. J Clean Prod 142:371–384. https://doi.org/10.1016/j.jclepro.2016.03.126

Guarnieri P, Cerqueira-Streit JA (2015) Implications for waste pickers of Distrito Federal, Brazil arising from the obligation of reverse logistics by the National Policy of Solid Waste. Lat Am J Manag Sustain Dev 2(1):19–35. https://doi.org/10.1504/LAJMSD.2015.067468

Guarnieri P, e Silva LC, Levino NA (2016) Analysis of electronic waste reverse logistics decisions using strategic options development analysis methodology: a Brazilian case. J Clean Prod 133:1105–1117. https://doi.org/10.1016/j.jclepro.2016.06.025

Guide VDR Jr, Van Wassenhove LN (2009) OR FORUM – the evolution of closed-loop supply chain research. Oper Res 57:10–18

Habuer, Nakatani J, Moriguchi Y (2014) Time-series product and substance flow analyses of end-of-life electrical and electronic equipment in China. Waste Manag 34:489–497. https://doi.org/10.1016/j.wasman.2013.11.004

Herat S, Agamuthu P (2012) E-waste: a problem or an opportunity? Review of issues, challenges and solutions in Asian countries. Waste Manag Res 30(11):1113–1129. https://doi.org/10.1177/0734242X12453378

Imran M, Haydar S, Kim J, Awan MR, Bhatti AA (2017) E-waste flows, resource recovery and improvement of legal framework in Pakistan. Resour Conserv Recycl 125:131–138. https://doi.org/10.1016/j.resconrec.2017.06.015

Iqbal M, Breivik K, Syed JH, Malik RN, Li J, Zhang G, Jones KC (2015) Emerging issue of e-waste in Pakistan: a review of status, research needs and data gaps. Environ Pollut 207:308–318. https://doi.org/10.1016/j.envpol.2015.09.002

Jiang L, Cheng Z, Zhang D, Song M, Wang Y, Luo C, Yin H, Li J, Zhang G (2017) The influence of e-waste recycling on the molecular ecological network of soil microbial communities in Pakistan and China. Environ Pollut 231(1):173–181. https://doi.org/10.1016/j.envpol.2017.08.003

Kamolkittiwong A, Phruksaphanrat B (2015) An analysis of drivers affecting green supply chain management implementation in electronics industry in Thailand. J Econ Bus Manag 3(9):864–869. https://doi.org/10.7763/JOEBM.2015.V3.299

Kilic HS, Cebeci U, Ayhan MB (2015) Reverse logistics system design for the waste of electrical and electronic equipment (E-WASTE) in Turkey. Resour Conserv Recycl 95:120–132. https://doi.org/10.1016/j.resconrec.2014.12.010

Kim S, Oguchi M, Yoshida A, Terazono A (2013) Estimating the amount of E-WASTE generated in South Korea by using the population balance model. Waste Manag 33:474–483. https://doi.org/10.1016/j.wasman.2012.07.011

Kumar A, Holuszko M, Espinosa DCR (2017) E-waste: an overview on generation, collection, legislation and recycling practices. Resour Conserv Recycl 122:32–42. https://doi.org/10.1016/j.resconrec.2017.01.018

Lagos G, Peters D, Videla A, Jara JJ (2018) The effect of mine aging on the evolution of environmental footprint indicators in the *Chilean* copper mining industry 2001–2015. J Clean Prod 174:389–400. https://doi.org/10.1016/j.jclepro.2017.10.290

Lepawsky J, Connolly CA (2016) Crack in the facade? Situating Singapore in global flows of electronic waste. Singap J Trop Geogr 37:158–175. https://doi.org/10.1111/sjtg.12149

Liu L, Zhang B, Lin K, Zhang Y, Xu X, Huo X (2018) Thyroid disruption and reduced mental development in children from an informal e-waste recycling area: a mediation analysis. Chemosphere 193:498–505. https://doi.org/10.1016/j.chemosphere.2017.11.059

Lombard R (2004) e-Waste assessment in South Africa. Case study of the Gauteng Province. Draft report

Nagurney A, Toyasaki F (2005) Reverse supply chain management and electronic waste recycling: a multitiered network equilibrium framework for e-cycling. Transp Res E 41(1):1–28. https://doi.org/10.1016/j.tre.2003.12.001

Nnorom IC, Osibanjo O (2008) Overview of electronic waste (e-waste) management practices and legislations, and their poor applications in the developing countries. Resour Conserv Recycl 52(6):843–858. https://doi.org/10.1016/j.resconrec.2008.01.004

Ongondo FO, Williams ID, Cherrett TJ (2011) How are E-WASTE doing? A global review of the management of electrical and electronic wastes. Waste Manag 31(4):714–730. https://doi.org/10.1016/j.wasman.2010.10.023

Osibanjo O, Nnorom IC (2007) The challenge of electronic waste (e-waste) management in developing countries. Waste Manag Res 25(6):489–501. https://doi.org/10.1177/0734242X07082028

Öztürk T (2015) Generation and management of electrical–electronic waste (e-waste) in Turkey. J Mater Cycles Waste Manag 17(3):411–421. https://doi.org/10.1007/s10163-014-0258-6

Pakistan Environmental Protection Act (PEPA) (1997)

Pakistan Import Policy Order (2016) Available at:. http://www.commerce.gov.pk/wp-content/uploads/pdf/IPO-2016.pdf. Accessed March 2018

Pakistan National Environment Policy (2005) Available at: https://www.mowr.gov.pk/wpcontent/uploads/2018/05/National-Environmental-Policy-2005.pdf. Accessed Aug 2017

Panambunan-Ferse M, Breiter A (2013) Assessing the side-effects of ICT development: E-waste production and management. A case study about cell phone end-of-life in Manado, Indonesia. Technol Soc 35(3):223–231. https://doi.org/10.1016/j.techsoc.2013.04.002

Park JE, Kang YY, Kim WI, Jeon TW, Shin SK, Jeong MJ, Kim JG (2014) Emission of polybrominated diphenyl ethers (PBDEs) in use of electric/electronic equipment and recycling of e-waste in Korea. Sci Total Environ 470:1414–1421. https://doi.org/10.1016/j.scitotenv.2013.07.129

Pascale A, Sosa A, Bares C, Battocletti A, Moll MJ, Pose D, Feola G (2016) E-waste informal recycling: an emerging source of Lead exposure in South America. Ann Glob Health 82(1):197–201. https://doi.org/10.1016/j.aogh.2016.01.016

Prakash C, Barua MK (2015) Integration of AHP-TOPSIS method for prioritizing the solutions of reverse logistics adoption to overcome its barriers under fuzzy environment. J Manuf Syst 37:599–615. https://doi.org/10.1016/j.jmsy.2015.03.001

Prakash C, Barua MK, Pandya KV (2015) Barriers analysis for reverse logistics implementation in Indian electronics industry using fuzzy analytic hierarchy process. Procedia Soc Behav Sci 189:91–102

Protomastro GF (2007) Estudio sobre los circuitos formales e informales de gestión de Residuos de Aparatos Eléctricos y Electrónicos en Argentina. 2007, e-srap, Ecogestionar-Ambiental del Sud SA: Buenos Aires

Protomastro G (2009) Electronic scrap management in Argentina. In: Lechner P (ed) Prosperity waste and waste resources. Proceedings of the 3rd BOKU waste conference, BOKU-University of Natural Resources and Applied Life Sciences, Vienna, Austria, April 15–17, pp 113–122

Ravi V, Shankar R (2005) Analysis of interactions among the barriers of reverse logistics. Technol Forecast Soc Chang 72(8):1011–1029. https://doi.org/10.1016/j.techfore.2004.07.002

REC Turkey (2011) Regulatory impact assessment of EU waste electrical and electronic equipment (E-WASTE) Directive (2002/96/EC)
Rhee SW (2016) Beneficial use practice of e-wastes in Republic of Korea. Procedia Environ Sci 31:707–714. https://doi.org/10.1016/j.proenv.2016.02.055
Rochman FF, Ashton WS, Wiharjo MGM (2017) E-waste, money and power: mapping electronic waste flows in Yogyakarta, Indonesia. Environ Dev 24(May 2016):1–8. https://doi.org/10.1016/j.envdev.2017.02.002
Rogers DS, Tibben-Lembke RS (1998) Going backwards: reverse logistics trends and practices. Reverse Logistics Executive Council, Reno
Sasaki S, Araki T (2013) Employer-employee and buyer-seller relationships among waste pickers at final disposal site in informal recycling: the case of Bantar Gebang in Indonesia. Habitat Int 40:51–57
Scruggs CE, Nimpuno N, Moore RBB (2016) Improving information flow on chemicals in electronic products and E-waste to minimize negative consequences for health and the environment. Resour Conserv Recycl 113:149–164. https://doi.org/10.1016/j.resconrec.2016.06.009
Shirodkar N, Terkar R (2017) Stepped recycling: the solution for E-waste management and sustainable manufacturing in India. Mater Today Proc 4(8):8911–8917. https://doi.org/10.1016/j.matpr.2017.07.242
Steubing B, Zah R, Waeger P, Ludwig C (2010) Bioenergy in Switzerland: assessing the domestic sustainable biomass potential. Renew Sust Energ Rev 14(8):2256–2265. https://doi.org/10.1016/j.rser.2010.03.036
Tang W, Cheng J, Zhao W, Wang W (2015) Mercury levels and estimated total daily intakes for children and adults from an electronic *waste* recycling area in Taizhou, China: key role of rice and fish consumption. J Environ Sci 34:107–115. https://doi.org/10.1016/j.jes.2015.01.029
United Nations Centre for Regional Development – UNCRD (2011) The International partnership for expanding waste management services in local authorities (IPLA). Available at: http://www.uncrd.or.jp/
United Nations Environment Programme – UNEP (2009) Recycling – from e-waste to resources. UNEP. Available at: http://www.greenbiz.com/sites/default/files/unep-ewaste-reoprt.pdf
Wang F, Kuehr R, Ahlquist D, Li J (2013) E-waste in China: a country report. UNU-ISP (Institute for Sustainability and Peace, United Nations University). Prepared for StEP (Solving the E-waste Problem)
Widmer R, Oswald-Krapf H, Sinha-Khetriwal D, Schnellmann M, Böni H (2005) Global perspectives on e-waste. Environ Impact Assess Rev 25(5 Special Issue):436–458. https://doi.org/10.1016/j.eiar.2005.04.001
Wong CSC, Duzgoren-Aydin NS, Aydin A, Wong MH (2007) Evidence of excessive releases of metals from primitive e-waste processing in Guiyu, China. Environ Pollut 148(1):62–72. https://doi.org/10.1016/j.envpol.2006.11.006
Xavier LH, Carvalho TCMB (2014) Introdução à Gestão de Resíduos de Equipamentos Eletroeletrônicos. In: Carvalho TCMB, Xavier LH (org) Gestão de Resíduos Eletroeletrônicos: uma abordagem prática para a sustentabilidade. Elsevier, Rio de Janeiro. (in Portuguese)
Yacob P, bin Mohamad Makmor MF, Zin AWBM, Aziz NSB (2012) Barriers to reverse logistics practices in Malaysian SMEs. Int J Acad Res Econ Manag Sci 1(5):204
Yadav V, Karmakar S, Dikshit AK, Vanjari S (2016) A feasibility study for the locations of waste transfer stations in urban centers: a case study on the city of Nashik, India. J Clean Prod 126:191–205. https://doi.org/10.1016/j.jclepro.2016.03.017
Yu J, Williams E, Ju M, Shao C (2010) Managing e-waste in China: policies, pilot projects and alternative approaches. Resour Conserv Recycl 54(11):991–999. https://doi.org/10.1016/j.resconrec.2010.02.006
Zhang S, Ding Y, Liu B, Chang C (2017) Supply and demand of some critical metals and present status of their recycling in E-WASTE. Waste Manag 65:113–127. https://doi.org/10.1016/j.wasman.2017.04.003

Chapter 6
Chemical Recycling of Electronic-Waste for Clean Fuel Production

Jayaseelan Arun and Kannappan Panchamoorthy Gopinath

Contents

Abstract Electronic-waste was the main waste stream raising concern to the researchers globally. Improper recycling and disposal techniques resulted in solemn effects on the atmosphere and public well-being. This chapter explains the systematic methods used for management of Electronic-waste. Electronic-waste managing would be an ideal start-up business platform toward energy production and metal recovery. The recycling pathways are designed by considering the current industrial reality and design strategies. Chemical recycling is a compilation of pyrolysis, cata-

J. Arun · K. P. Gopinath (✉)
Department of Chemical Engineering, SSN College of Engineering,
Kalavakkam, Tamil Nadu, India
e-mail: gopinathkp@ssn.edu.in

A. Khan et al. (eds.), *E-waste Recycling and Management*,
Environmental Chemistry for a Sustainable World 33,
https://doi.org/10.1007/978-3-030-14184-4_6

lytic cracking/upgrading, gasification, and chemolysis methods. Pyrolyzing of Electronic-waste prior to catalytic cracking method yielded high-quality oil. This oil can be further upgraded into clean fuels. Integrated process (pyrolysis and catalytic upgrading) results in considerable financial and ecological benefits during processing Electronic-waste into clean fuels.

Keywords Electronic-waste · chemical recycling · Clean fuel · Energy · Valuable chemical · Plastics · Hydrothermal · Gasification · Combustion · Environment

6.1 Introduction

Modernized electronic inventions with shorter lifespan of electronic goods made electronic commerce as the primary budding sector globally. Subsequently, this paved way for the generation of waste electrical and electronic equipments in huge quantity annually. Well-developed and budding countries are facing a serious challenge in Electronic-waste management. Waste electrical and electronic equipments have received attractions universally because of its unique characteristics, energy value, and impact on the environment and individual healthiness (Ongondo et al. 2011; Perez-Belis et al. 2015). Electronic-waste generation was growing at an exponential rate as much as three times higher than a municipal waste generation (Rahmani et al. 2014). Researchers on recycling plastic wastes have reached a significant level taking into consideration the ecological benefits and energy demand of the society.

Electronics and electrical wastes are inhomogeneous and composite in terms of composition and equipment makeup. They are toxic as heavy metals and hence need safe usage and recycling to keep away from destructive effects on human and environment well-being (Freegard et al. 2006; Song and Li 2015). Various materials can be recovered from waste electrical and electronic equipment recycling process (Widmer et al. 2005). Currently, four methods available for treating waste electrical and electronic equipment plastics were landfilling, incineration, mechanical recycling, and chemical recovery. Apart from these four methods, pyrolysis is also adopted in many countries for producing hydrocarbons and chemical compounds.

Electronic-waste recycling due to poor technical capacity and inadequate collection methods have achieved only 13% recycling rate despite various recycling technologies available throughout the world (Jiang et al. 2012). Globally research on Electronic-waste recycling was still far from generating closed loop systems for efficient processing of Electronic-waste to recover valuable chemicals (Li et al. 2015). Theoretical guide to recycling waste should make use of the precedent experience, and it should deal with the current electronics production rate. The eco-friendly design attracted the consumers, recyclers, and manufacturers (Stevels et al. 2013).

6.2 Electronic-Waste: A Business Platform

Electronic-waste or waste electrical and electronic equipment was generated from a vast and broad range of household equipments (refrigerators, mobile phones, air conditioners, etc.) and electronic computers which are discarded by their owners (Nnorom and Osibanjo 2008a). Figure 6.1 elaborates the Electronic-waste production globally from various goods. Waste electrical and electronic equipments mainly comprised of ferrous and nonferrous metals and plastics (Fig. 6.2) (Huisman et al. 2008). Electronic-waste produced in 2014 possessed 16.5 million tons (Mt) of iron, 1.9 Mt of copper, and 8.6 Mt of plastics around a predictable cost of US $52 billion (Balde et al. 2015). Additionally, toxic materials consist of lead glass (2.2 Mt), batteries (0.3 Mt), and 4400 t of ozone-depleting substances like mercury, cadmium, and chromium. Due to these compositions, waste electrical and electronic equipments emerge as a secondary resource.

Electronic-waste was the resource of income for industries as well as offers new jobs for the public. In India, Bangalore produces 18,000 metric tons of Electronic-waste annually. Table 6.1 elaborates the various Electronic-wastes types and their source of equipments which they are generated. Presence of gold, platinum, aluminum, copper, and earth metals in Electronic-waste is enough to reuse and provides a huge turn over for the industries. Plastic content in Electronic-waste was a good raw material for pyrolysis process and thermochemical treatment process. Pyrolysis oil recovered after decomposing Electronic-waste was used as a substitute for diesel in generators.

6.2.1 Plastics in Electronic-Waste

Plastics are the derivatives of petrochemicals which are derived from petroleum fuels (OIL 2008). Plastics are the synthetic resources comprising of macromolecular composites; they can be recycled back to their raw assets under appropriate

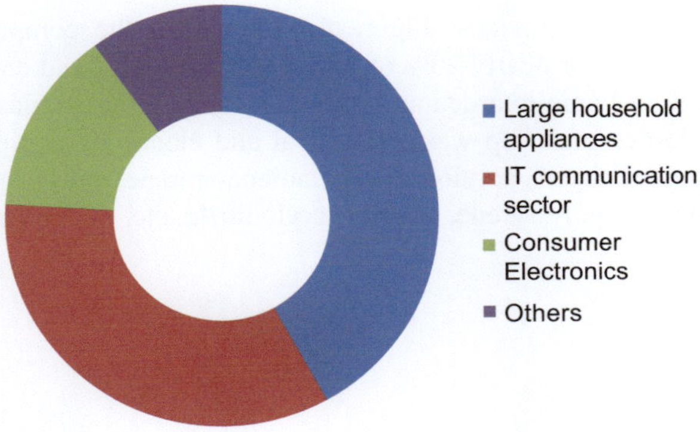

Fig. 6.1 Various sectors responsible for Electronic-waste generation

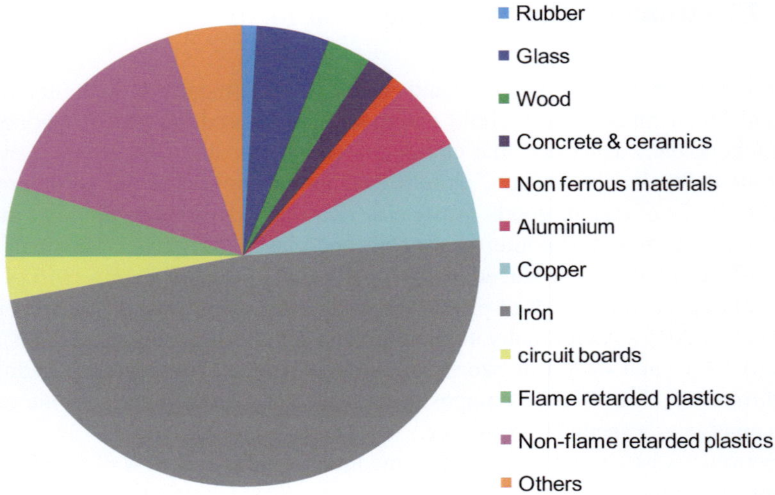

Fig. 6.2 The composition of waste electrical and electronic equipment

Table 6.1 Electronic-waste and its source of electronic instruments

S. no	Waste category	Source of Electronic-waste
1	IT and telecommunication	LAN, cell phones, printers, modems
2	Gadgets	MP3 players, DVD players, digital camera, computers
3	Major household products	Air conditioners, refrigerator, washing machine, micro-oven
4	Minor household products	Video game consoles, electric kettles, television, grinder
5	Electrical and electronic products	Transistors, diodes, integrated circuits, batteries, transformers, resistors, wires
6	Monitoring instruments	Thermostat, microcontrollers, relays
7	Medical devices	Biomedical instruments, thermometer
8	Automated instruments	Automatic soap and water dispensers, etc.

process parameters. Figure 6.2 elaborates the composition of plastics. Plastics account for nearly 20% of total waste electrical and electronic equipment plastics, 5% of flame-retarded and 15% non-flame-retarded plastics. More than 15 types of plastics make up waste electrical and electronic equipments, namely, polyesters, polyurethane, acrylonitrile-butadiene-styrene, polyethylene, polyamide, polypropylene, polystyrene, styrene acrylonitrile, etc. (Vilaplana and Karlsson 2008).

6.2.2 Electronic-Waste Management Issues

Technical innovation and economic development in developing countries made them produce a huge amount of electrical and electronic equipments (Hossain et al. 2015). Also Electronic-waste management was the huge headache for countries, since they are produced or imported as used items (Nnorom and Osibanjo 2008b). Due to low income and middle income, Electronic-waste is disposed of in unsanitary landfill sites. Wires are burned down to recover copper in it. Printed circuit boards are acid extracted to retrieve gold, platinum, palladium, and silver coated in them. These kinds of activities are seen in developing countries like India, China, Pakistan, Nigeria, and Ghana; somewhere they lack the facility to safeguard health and environment (Leung et al. 2006; SEPA 2011). Seitz (2014) revealed Electronic-waste was a rising concern among developing countries toward the environment and public health.

6.2.3 Worldwide Electronic-Waste Generation

Fossil fuels were exploited heavily and utilized as a cheap energy source. If these exist, they get depleted in near future which paves way for the emerging of secondary energy source. In 2014 worldwide 41.8 Mt of Electronic-waste was produced, and it was estimated to increase around 50 Mt in 2018 at a yearly growth rate of 5% (Balde et al. 2015). China is an emerging economy and the largest electronic-manufacturing country which makes them second in waste electrical and electronic equipment generation next to the USA (McCann and Wittmann 2015). Emerging countries were generating twice the amount of Electronic-waste than the urbanized countries. Developed countries were even depositing their Electronic-waste onto the developing countries leading to serious issues. This causes serious environmental problems and health issues for the local population. Table 6.2 elaborates the types of waste produced across various developing and developed countries.

Table 6.2 Literature review on types of waste produced across countries

Country	Type of waste	References
Jordan	Electronic-waste	Ikhlayel (2017)
Iran	MSW	Abduli et al. (2011)
Italy	MSW	Buratti et al. (2015)
Brazil	Electronic-waste	De-Souza et al. (2016)
China	Electronic-waste	Hong et al. (2015)
Jordan	MSW	Ikhlayel et al. (2016)
Vietnam	MSW	Thanh and Matsui (2013)
Macau	Electronic-waste	Song et al. (2013)
China	Electronic-waste	Bian et al. (2016)
Sakarya	MSW	Erses-Yay (2015)

6.2.4 Electronic-Waste on Environmental Public Health

Several types of research were under progress to study the effect of Electronic-waste on the environment. Xue et al. (2015) reported the impact of formal recycling of printed circuit boards on the environment. Fujimori et al. (2012) reported the enhancement factors, dangerous indicators, and concentration of metals present in soil due to proper and improper recycling of Electronic-waste. Mostly, major studies are carried on focusing on the emissions from improper recycling of wastes. Some studies are conducted on assessing the effect of Electronic-waste on health.

6.3 Energy Recovery from Electronic-Waste

Plastic wastes can be incinerated in bulk quantity to generate energy; due to the presence of high-value polymers, they can act as an alternative fuel resource. However, energy recovery may be ecologically resourceful in the case of bulk handling of plastics by fulfilling the emission regulations and energy need. Figure 6.3 elaborates the methodology for energy recovery from Electronic-waste.

Recycling of plastics through mechanical and chemical methods is getting its importance than land filling and incineration methods. Chemical recycling is an economically feasible technique for waste electrical and electronic equipment treatment, including methods like pyrolysis, hydrothermal treatment, and catalytic pyrolysis toward converting waste electrical and electronic equipment plastics into chemicals and high-energy fuels.

6.3.1 Chemical Recycling

Plastic waste and Electronic-waste are used as feedstock for generation of fuels and valuable products. Globally the interest was not only on treatment of waste but also on recovery of some eco-friendly products like petrochemical feedstock. These feedstocks possess higher hydrocarbon content than other biomass. Base material is cheaper than the chemically recycled polymers due to capital investment, raw material cost, etc. Polyethylene terephthalate methanolysis was carried out with methanol under higher temperatures (180–280 °C) and pressures (20–40 atm), yielding dimethyl terephthalate and ethylene glycol. Table 6.3 describes the advantages and challenges of the chemical and mechanical recycling process. Matsushita Electric Works, Ltd., Japan, was generating a depolymerization methodology for treating flame-retardant polymers through hydrolysis under subcritical water. In this methodology, recycling rate of 70% was achieved on recycling thermosetting resin in flame-retardant polymers into basic materials.

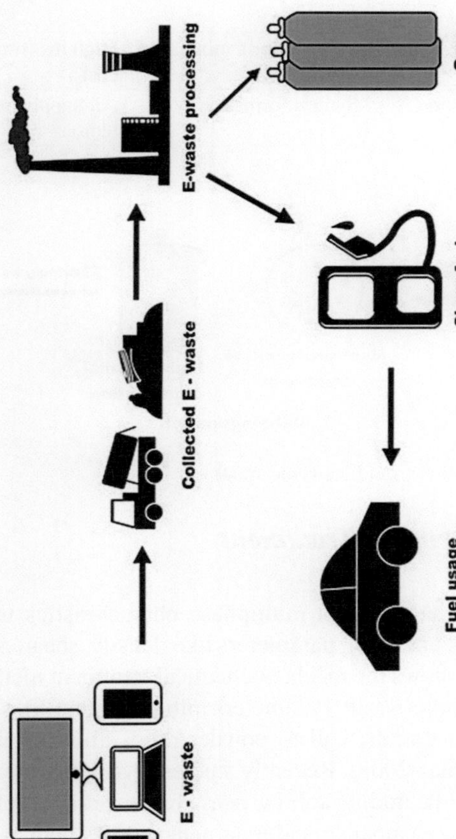

Fig. 6.3 Energy recovery from Electronic-waste

Table 6.3 Advantages and demerits of chemical recycling and mechanical recycling process

	Technique	Advantages	Demerits
Mechanical recycling	Flotation/sorting	Cheap	Restricted to two mixtures
	Reprocessing	Higher recycling value	The thermal mechanical decomposition process
Chemical recycling	Chemolysis	Synthesis of value-added chemicals	Bulk processing makes cost effective
	Pyrolysis	Easy to handle	Low tolerance to PVC
	Catalytic cracking	Narrow product formation	The catalyst cannot be reused
	Hydrocracking	Suitable for a mixture of plastics	High investment and operational cost
	Gasification	Syngas formation	Air supply and processing is tedious

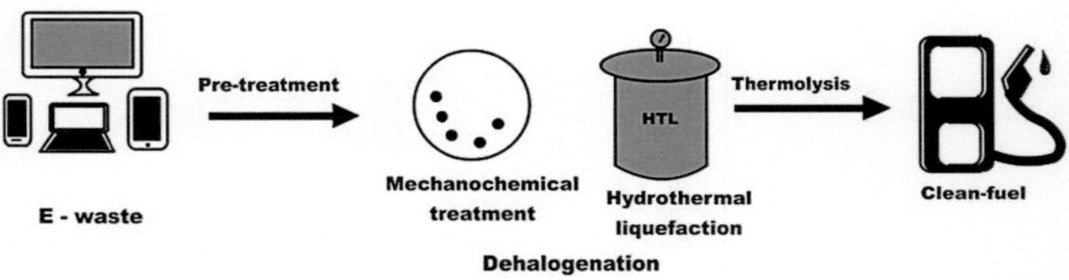

Fig. 6.4 Clean fuel production from Electronic-waste

6.3.2 Mechanochemical Treatment

Mechanochemistry-mechanical and multiphase characteristics need higher energized mills with various operating parameters like density, shear, and impact (Balaz et al. 2013). Figure 6.4 shows the mechanochemical treatment method for clean fuel production from Electronic-waste. Parameters influencing milling practice were the mill type, processing materials, ball to powder ratio, filling chamber, processing speed and time, etc. (Balaz 2008). Recently studies reveal that the mechanochemistry method was used for degrading solid wastes and recycling of plastics (Guo et al. 2010). Advantages of mechanochemical treatment process than conventional methods were an uncomplicated process, ecological safety, and product that can be recovered in a meta-stable state. Acrylonitrile-butadiene-styrene polymers and high-impact polystyrene and styrene copolymers are the various types of engineering polymers present in polymeric composites (16 wt% of waste electrical and electronic equipments). Thermochemical recycling of plastics yields monomers and fuel under pyrolytic conditions (Grause et al. 2011).

A new methodology of bromine removal from styrene polymers was reported in literature (Grause et al. 2015). Decabromodiphenyl ethane was removed effectively through NaOH/ethylene glycol solution in moderate environment (150–190 °C) in a ball mill reactor. Debromination is done by substituting hydroxide or elimination of hydrogen bromide. Once decabromodiphenyl ethane is removed, the residue is suitable for mechanical recycling. After the removal of brominated organic compounds, the quality of fuel products is upgraded and their environmental effects are decreased.

Dehalogenation of solid plastics is increased by adding additives during mechanochemical treatment process. Generally, additives like alkali metal oxides (calcium oxide, sodium hydroxide), iron powders, and quartz (silica) are employed as catalytic adsorbents. Since, these materials are unsustainable, novel method was developed through simultaneous grinding of plastics in the midst of sustainable resources (e.g., biopolymers, biowastes, eco-friendly minerals) through mechanochemical treatment for clean fuel synthesis was elaborated in Fig. 6.4.

6.3.3 Hydrothermal Process

Thermal conduct of waste electrical and electronic equipment plastics emerged as a suitable skill by degrading organo-bromine compounds, also in situ and secure exclusion of bromine constituents from the oil products. The supercritical fluid technology has emerged as a potential technique for chemical recycling of plastic wastes. Supercritical fluids act as a better chemical medium under optimum conditions for depolymerization, hydrolysis, hydrogenation, and dehydrogenation with properties like low viscosity, low dielectric content, high mass transport coefficient, and higher diffusivity (Shibasaki et al. 2004; Zhang et al. 2013).

Hydrothermal treatment was preferred for clean fuel production because of its higher efficiency than pyrolysis of biomass and sewage waste (Yu et al. 2016; Shen et al. 2016). Reactor corrosion and higher energy utilization is the only drawback in supercritical fluid method (Guo et al. 2009). Selection of appropriate supercritical fluid and enhancers, price and operating parameters, etc. are the common challenges faced during the hydrothermal treatment process.

Hydrothermal process is of two major types: (i) hydrothermal liquefaction and (ii) hydrothermal gasification. The quality of final product is decided by the operating parameters and environment (Yan et al. 2010). Dehalogenation by hydrothermal treatment method has been studied in recent days for plastics compounds (Starnes 2012). Solid fuel properties are significantly increased by unification of biomass through hydrothermal conditions.

6.3.4 Pyrolysis

Pyrolysis is an environment-friendly and economically feasible technique for waste electrical and electronic equipment plastic treatment than landfilling and incineration. Emission of toxic gases into the environment is lesser than incineration process

(Bhaskar et al. 2002). Pyrolysis is an efficient method of valorizing Electronic-waste which can recuperate the valuable compounds with low emission of pollutants into the atmosphere. Plastic wastes are converted into fuel by fast pyrolysis since it is a promising technique for protecting the environment from these nondegradable plastics. During pyrolysis, plastics are thermally degraded and rehabilitated into oil, gaseous, and charred products at (700–900 K) in an inert atmosphere. Pyrolysis generates bio-oil with high bromine and chlorine content (Lopez et al. 2011).

Pyrolysis was carried out as single-step cracking process under the closed environment. Fixed, fluidized bed and tubular reactors are used for pyrolysis process and divided into three major processes based on process parameters like conventional, fast, and slow pyrolysis (Wu and Williams 2013). Fluidized bed reactor acts as a better heat and mass transfer equipment yielding thin-layered plastic, suggesting the polymer degradation. Fast and slow pyrolysis methods follow the finest route for converting brominated flame-retardant plastics onto clean fuels and valuable products than conventional pyrolysis process. Recently, a study revealed that printed circuit boards pyrolysis resulted in a higher content of bromine, glass fibers, and metals (Copper) (Shen et al. 2018).

6.3.4.1 Thermal Pyrolysis

Waste when subjected to thermal decomposition under zero oxygen environment yields char, oil, and gaseous products which are further upgraded and used as fuels. Several studies were reported on pyrolysis of brominated flame-retardant plastics in various reactors (Hall and Williams 2006; Jung et al. 2012; Miskolczi et al. 2008). Results suggest that pyrolysis of brominated high-impact polystyrene resulted in higher yield of oil in fixed bed reactor. Pyrolysis oil contains toluene, ethylbenzene, styrene, and cumene. Pyrolysis of brominated high-impact polystyrenes produced 98 wt% of oil containing 61.7 wt% of volatile products which were resulted due to the thermal steadiness of polymeric chains. Pyrolysis of brominated high-impact polystyrenes yielded oil around 500 mg/g plastic. In contrast, brominated acrylonitrile-butadiene-styrene pyrolysis yields 400 mg/g of plastic.

6.3.4.2 Co-pyrolysis

Combined pyrolysis technique basically deals with two or more dissimilar materials as resource to yield oil with improved quality and quantity. Co-pyrolysis can reduce manufacturing cost and resolve several problems in managing waste. Due to inherent complexity, several issues arise in Electronic-waste management. Co-pyrolysis improves quality and quantity of pyrolysis oil without any catalyst or solvents, which made this method an unavoidable technique in industrial applications (Abnisa and Daud 2014).

6.3.5 Combustion Process

Combustion of fossil fuels was replaced by biomass and wastes for energy and heat generation. This process is a technically feasible method to reduce harmful greenhouse gases (carbon dioxide) into the environment. However, replacing conventional fossil fuels ended up generating a huge amount of ash-related problems (slagging, corrosion, and fouling). Alkali metal usage can overcome these problems (Hansen et al. 2000). Brominated fuel possesses a promising effect on volatilization of metals like potassium, iron, copper, zinc, and lead (Vehlow et al. 2003). Halogen hydrides and small-chain halogenic organic compounds were resulted from decomposition of organic halogenated compounds. Chlorinated plastics (waste electrical and electronic equipments, polyvinyl chloride, textiles) and halogen hybrids (hydrogen chloride, hydrogen bromide) were the chief products produced through combustion method (Wu et al. 2014).

Brominated flame retardants containing wastes generate polybrominated dibenzo-p-dioxins and polybrominated dibenzofurans through the course of combustion process (Wang and Zhang 2012). Under thermal conditions they are involved in the recycling process. Polybrominated diphenyl ethers act as a substrate for production. Insufficient combustion process or disturbed process leads to fire accidents, uncontrolled burning, and gasification.

6.3.6 Gasification Process

Pyrolysis under elevated temperature generates fuels (oil, gas) possessing higher heating value. Liquid fuels produced from circuit boards through pyrolysis at 800 °C in static temperature conditions possessed brominated compounds; this made them unusable without further downstream processing (William and Paul 2007). Partial oxidation of waste electrical and electronic equipment plastic at an elevated temperature (1200 °C) decreased the brominated or chlorinated dioxins in gas products. Nevertheless the halogen compounds in gaseous products were not in permissible limits for use as fuels. Majority of organo-brominated compounds in brominated flame retardants are broken down into hydrogen bromide and bromine at higher temperature due to their fundamentals (Jin et al. 2011). Usage of calcium oxide deliberately increases the inorganic bromine formation from the organic bromine compounds. Burning circuit boards at elevated temperature effectively breaks down organo-brominated compounds.

Steam gasification emerges as a promising technique because of using carbonate in the recycling of waste electrical and electronic equipment plastics. Halogenated compounds present within waste electrical and electronic equipment plastics were retrieved in the form of stable organic salts (Zhang et al. 2013). Lithium carbonate, sodium carbonate, and potassium carbonate are used as a catalyst in steam gasification under mild conditions. During steam gasification the carbonate or biomass cannot account for the halogen emission but accelerates the transformation of tar and char into gas products from plastics (Lopez et al. 2015).

6.3.7 Integrated Process

Numerous methods like mechanochemical treatment and hydrothermal treatment processes were employed to remove halogenated compounds present in plastics. In that case the solid waste was upgraded through hydrothermal treatment process. Table 6.4 elaborates the merits and demerits of dehalogenation process through waste plastic recycling. Mechanochemical treatment or hydrothermal treatment process is uncomplicated and eco-friendly. Energy utilization was higher since solid plastics are processed for energy (fuel, oil, etc.) generation. The downstream process needs to be established for these treatment processes. Through sorption and dehalogenation process, only the produced polybrominated diphenyl ethers and printed boards were removed (Huang et al. 2013; Zhuang et al. 2011). Low cost and sustainable additive addition during hydrothermal treatment or mechanochemical treatment process may end up in synergistic effects in thermal applications. In view of industries, mechanochemical treatment and hydrothermal treatment process integrated with catalytic thermal degradation is the ideal method to produce clean fuels from Electronic-waste.

6.3.8 Hydrocracking

Hydrocracking process differs from catalytic upgrading of solid plastics only by hydrogen usage in it. This method was carried out at 70 atm and temperature of 375–400 °C in the presence of a catalyst. Hydrogen usage enhanced the final product quality (superior H/C ratio along with lesser aromatic compounds). The mixture of plastics can be hydrocracked to produce a good quality of naphtha. But this process needs a higher-operating pressure and investment cost.

Table 6.4 Merits and demerits of plastic wastes dehalogenation

Method	Advantages	Disadvantages
Pyrolysis	Simple process, low energy consumption and produce high esteemed products (fuels)	Halogenated compounds (dioxin) can form, costly process
Gasification	Easy process, produce syngas and fuels, upgrading of these products is easier	Halogenated compounds (dioxin) can form, plastic wastes are not completely degraded
Combustion	Simple process, complete degradation of plastic wastes	A lesser amount of halogenated compounds (dioxin) forms, high energy consumption
Hydrothermal	Higher efficiency, time-saving	High cost, energy consumption high
Mechanochemical	Simple process, need additives addition	Lesser efficiency, high energy consumption

6.4 Conclusion

Waste management sector needs an integrative thinking and innovative ideas to solve the issues in modern societies. Electronic-waste generated was kept for a shorter span due to lack of safe discarding procedures and improper recycling facilities. Pre-treatment through mechanochemical treatment or hydrothermal treatment process eliminates halogens in plastic wastes. Dehalogenation of plastic wastes was obtained through combined grinding of sustainable wastes. Dehalogenations through mechanochemical treatment or hydrothermal treatment methods are handy as well as environmental process. In view of industries, incorporated method of mechanochemical treatment or hydrothermal treatment with the catalytic thermal methods was preferred for the clean fuel making from Electronic-waste.

References

Abduli MA, Naghib A, Yonesi M, Akbari A (2011) Life cycle assessment (LCA) of solid waste management strategies in Tehran: landfill and composting plus landfill. Environ Monit Assess 178:487–498

Abnisa F, Daud WMAW (2014) A review on co-pyrolysis of biomass: an optional technique to obtain high-grade pyrolysis oil. Energy Convers Manag 87:71–85

Balaz P (2008) Mechanochemistry in nanoscience and minerals engineering. Springer, Berlin/Heidelberg

Balaz P, Achimovieova M, Balaz M, Billik P, Cherkezova-Zheleva Z, Criado JM et al (2013) Hallmarks of mechanochemistry: from nanoparticles to technology. Chem Soc Rev 42:7571

Balde C, Wang F, Kuehr R, Huisman J (2015) The global electronic waste monitor. United Nations University, IAS–SCYCLE, Bonn

Bhaskar T, Matsui T, Kaneko J, Uddin MA, Muto A, Sakata Y (2002) Novel calcium based sorbent (Ca-C) for the dehalogenation (Br, Cl) process during halogenated mixed plastic (PP/PE/PS/PVC and HIPS-Br) pyrolysis. Green Chem 4:372–375

Bian J, Bai H, Li W, Yin J, Xu H (2016) Comparative environmental life cycle assessment of waste mobile phone recycling in China. J Clean Prod 131:209–218

British Plastics Federation (2008) Oil consumption: what happens to plastics when the oil runs out and when will it run out. Available from: http://www.bpf.co.uk/press/Oil_Consumption.aspx

Buratti C, Barbanera M, Testarmata F, Fantozzi F (2015) Life cycle assessment of organic waste management strategies: an Italian case study. J Clean Prod 89:125–136

De-Souza RG, Climaco JCN, Sant'Anna AP, Rocha TB, do Valle RDB, Quelhas OLG (2016) Sustainability assessment and prioritisation of electronic waste management options in Brazil. Waste Manag 57:46–56

Erses-Yay AS (2015) Application of life cycle assessment (LCA) for municipal solid waste management: a case study of Sakarya. J Clean Prod 94:284–293

Freegard K, Tan G, Coggins-Wamtech C, Environmental DFD, Alger M, Cracknell P et al (2006) Develop a process to separate brominated flame retardants from WASTE ELECTRICAL AND ELECTRONIC EQUIPMENTS polymers. (Final Report). The Waste & Resources Action Programme, London

Fujimori T, Takigami H, Agusa T, Eguchi A, Bekki K, Yoshida A, Terazono A, Ballesteros FC (2012) Impact of metals in surface matrices from formal and informal electronic-waste recycling around Metro Manila, the Philippines, and intra-Asian comparison. J Hazard Mater 221:139–146

Grause G, Karakita D, Ishibashi J, Kameda T, Bhaskar T, Yoshioka T (2011) TGMS investigation of brominated products from the degradation of brominated flame retardants in high-impact polystyrene. Chemosphere 85:368–373

Grause G, Fonseca JD, Tanaka H, Bhaskar T, Kameda T, Yoshioka T (2015) A novel process for the removal of bromine from styrene polymers containing brominated flame retardant. Polym Degrad Stab 112:86–93

Guo J, Guo J, Xu ZM (2009) Recycling of non-metallic fractions from waste printed circuit boards: a review. J Hazard Mater 168:567–590

Guo X, Xiang D, Duan G, Mou P (2010) A review of mechanochemistry applications in waste management. Waste Manag 30:4–10

Hall WJ, Williams PT (2006) Pyrolysis of brominated feedstock plastic in a fluidised bed reactor. J Anal Appl Pyrolysis 77:75–82

Hansen LA, Nielsen HP, Frandsen FJ, Dam-Johansen K, Hørlyck S, Karlsson A (2000) Influence of deposit formation on corrosion at a straw-fired boiler. Fuel Process Technol 64:189–209

Hong J, Shi W, Wang Y, Chen W, Li X (2015) Life cycle assessment of electronic waste treatment. Waste Manag 38:357–365

Hossain M, Al-Hamadani S, Rahman R (2015) Electronic waste: a challenge for sustainable development. J Health Pollut 5:550–555

Huang Q, Liu W, Peng P, Huang W (2013) Reductive debromination of tetra bromobisphenol a by Pd/Fe bimetallic catalysts. Chemosphere 92:1321–1327

Huisman J, Magalini F, Kuehr R, Maurer C, Ogilvie S, Poll J et al (2008) Review of directive 2002/96 on waste electrical and electronic equipment (WASTE ELECTRICAL AND ELECTRONIC EQUIPMENTS). United Nations University, Bonn

Ikhlayel M (2017) Environmental impacts and benefits of state-of-the-art technologies for electronic waste management. Waste Manag 68:458–474

Ikhlayel M, Higano Y, Yabar H, Mizunoya T (2016) Introducing an integrated municipal solid waste management system: assessment in Jordan. J Sustain Dev 9:43

Jiang P, Harney M, Song Y, Chen B, Chen Q, Chen T, Lazarus G, Dubois LH, Korzenski MB (2012) Improving the end-of-life for electronic materials via sustainable recycling methods. Procedia Environ Sci 16:485–490

Jin Y, Tao L, Chi Y, Yan J (2011) Conversion of bromine during thermal decomposition of printed circuit boards at high temperature. J Hazard Mater 186:707–712

Jung SH, Kim SJ, Kim JS (2012) Thermal degradation of acrylonitrile–butadiene– styrene (ABS) containing flame retardants using a fluidized bed reactor: the effects of ca-based additives on halogen removal. Fuel Process Technol 96:265–270

Leung A, Cai ZW, Wong MH (2006) Environmental contamination from electronic waste recycling at Guiyu, Southeast China. J Mater Cycles Waste Manage 8:21–33

Li J, Zeng X, Chen M, Ogunseitan OA, Stevels A (2015) "Control-Alt-Delete": rebooting solutions for the electronic waste problem. Environ Sci Technol 49:7095–7108

Lopez A, de Marco I, Caballero BM, Laresgoiti MF, Adrados A (2011) Dechlorination of fuels in pyrolysis of PVC containing plastic wastes. Fuel Process Technol 92:253–260

Lopez G, Ekiaga A, Amutio M, Bilbao J, Olazar M (2015) Effect of polyethylene co-feeding in the steam gasification of biomass in a conical spouted bed reactor. Fuel 153:393–401

McCann D, Wittmann A (2015) Solving the electronic waste problem (Step) green paper: E- waste Prevention, take-back system design and policy approaches. United Nations University/Step Initiative, Germany

Miskolczi N, Hall WJ, Angyal A, Bartha L, Williams PT (2008) Production of oil with low organobromine content from the pyrolysis of flame retarded HIPS and ABS plastics. J Anal Appl Pyrolsis 83:115–123

Nnorom IC, Osibanjo O (2008a) Overview of electronic waste (electronic waste) management practices and legislations, and their poor applications in the developing countries. Resour Conserv Recycl 52:843–858

Nnorom IC, Osibanjo O (2008b) Sound management of brominated flame retarded (BFR) plastics from electronic wastes: state of the art and options in Nigeria. Resour Conserv Recycl 52:1362–1372

Ongondo FO, Williams ID, Cherrett TJ (2011) How are WASTE ELECTRICAL AND ELECTRONIC EQUIPMENTS doing? A global review of the management of electrical and electronic wastes. Waste Manag 31:714–730

Perez-Belis V, Bovea M, Ibanez-Fores V (2015) An in-depth literature review of the waste electrical and electronic equipment context: trends and evolution. Waste Manag Res 33:3–29

Rahmani M, Nabizadeh R, Yaghmaeian K, Mahvi AH, Yunesian M (2014) Estimation of waste from computers and mobile phones in Iran. Resour Conserv Recycl 87:21–29

Seitz J (2014) Analysis of existing electronic waste practices in MENA countries. The regional solid waste exchange of information and expertise network in Mashreq and Maghreb Countries (SWEEP-Net)

SEPA (2011) Recycling and disposal of electronic waste: health hazards and environmental impacts. Naturvårdsverket, Stockholm

Shen Y, Zhao R, Wang J, Chen X, Ge X, Chen M (2016) Waste-to-energy: De-halogenation of plastic-containing wastes. Waste Manag 49:287–303

Shen Y, Chen X, Ge X, Chen M (2018) Chemical pyrolysis of electronic waste plastics: char characterization. J Environ Manag 214:94–103

Shibasaki Y, Kamimori T, Kadokawa J, Hatano B, Tagaya H (2004) Decomposition reactions of plastic model compounds in sub and supercritical water. Polym Degrad Stab 83:481–485

Song Q, Li J (2015) A review on human health consequences of metals exposure to electronic waste in China. Environ Pollut 196:450–461

Song QB, Wang ZS, Li JH (2013) Sustainability evaluation of electronic waste treatment based on emergy analysis and the LCA method: a case study of a trial project in Macau. Ecol Indic 30:138–147

Starnes WH (2012) How and to what extent are free radicals involved in the nonoxidative thermal dehydrochlorination of poly (vinyl chloride)? J Vinyl Addit Technol 18:71–75

Stevels A, Huisman J, Wang F, Li J, Li B, Duan H (2013) Take back and treatment of discarded electronics: a scientific update. Front Environ Sci Eng 7:475–482

Thanh NP, Matsui Y (2013) Assessment of potential impacts of municipal solid waste treatment alternatives by using life cycle approach: a case study in Vietnam. Environ Monit Assess 185:7993–8004

Vehlow J, Bergfeldt B, Hunsinger H, Seifert H, Mark FE (2003) Bromine in waste incineration: partitioning and influence on metal volatilisation. Environ Sci Pollut Res Int 10:329–334

Vilaplana F, Karlsson S (2008) Quality concepts for the improved use of recycled polymeric materials: a review. Macromol Mater Eng 293:274–297

Wang Y, Zhang FS (2012) Degradation of brominated flame retardant in computer housing plastic by supercritical fluids. J Hazard Mater 205–206:156–163

Widmer R, Oswald-Krapf H, Sinha-Khetriwal D, Schnellmann M, Boni H (2005) Global perspectives on electronic waste. Environ Impact Assess Rev 25:436–458

William JH, Paul TW (2007) Separation and recovery of materials from scrap printed circuit boards. Res Conserv Recycl 51:691–709

Wu C, Williams PT (2013) Advanced thermal treatment of wastes for fuels, chemicals and materials recovery. In: Hester RE, Harrison RM (eds) Waste as a resource. The Royal Society of Chemistry, Cambridge, pp 1–43

Wu H, Shen Y, Harada N, An Q, Yoshikawa K (2014) Production of pyrolysis oil with low bromine and antimony contents from plastic material containing brominated flame retardants and antimony trioxide. Energy Environ Res 4:105–118

Xue M, Kendall A, Xu Z, Schoenung JM (2015) Waste management of printed wiring boards: a life cycle assessment of the metals recycling chain from liberation through refining. Environ Sci Technol 49:940–947

Yan W, Hastings JT, Acharjee TC, Coronella CJ, Vásquez VR (2010) Mass and energy balances of wet torrefaction of lignocellulosic biomass. Energy Fuel 24:4738–4742

Yu J, Sun L, Ma C, Qiao Y, Yao H (2016) Thermal degradation of PVC: a review. Waste Manag 48:300–314

Zhang S, Yoshikawa K, Nakagome H, Kamo T (2013) Kinetics of the steam gasification of a phenolic circuit board in the presence of carbonates. Appl Energy 101:815–821

Zhuang Y, Ahn S, Seyfferth AL, Masue-Slowey Y, Fendorf S, Luthy RG (2011) Dehalogenation of polybrominated diphenyl ethers and polychlorinated biphenyl by bimetallic, impregnated, and nanoscale zerovalent iron. Environ Sci Technol 45:4896–4903

Chapter 7
Management of Waste Electrical and Electronic Equipment in European Union Countries: A Comparison

Isabel Narbón-Perpiñá and Diego Prior

Contents

Abstract Over the last decades, the increasing volume of waste electrical and electronic equipment (WEEE) has become a major matter of concern worldwide. In the European case, a specific legislation has been developed in order to address the environmental problems associated with the proper management of this particular form of hazardous waste (Directive 2012/19/EU). Accordingly, it has introduced specific targets for the reuse and recycling and recovery of WEEE, which European countries should include in their national policies.

The aim of this chapter is to contribute to the literature on the management of WEEE by comparing the performance of the different European Union countries according to the targets set in the regulation of the Union's environmental policy on WEEE. To this end, we use the traditional nonparametric Data Envelopment Analysis (DEA) in order to measure technical efficiency for the first time in the literature. We use a sample of 30 European countries for the year 2014, with the purpose of comparing their performance, ranking the countries, and identifying their level of inefficiency. Our results suggest that European countries are highly efficient in the implementation of the WEEE Directive and management of the recycling and recovery of WEEE. However, considering the performance of different waste categories, we observe significant differences in WEEE efficiency among countries. Specifically, more efforts are needed to achieve higher efficiency levels from small equipment, lights, electrical and electronic tools, and medical devices.

I. Narbón-Perpiñá · D. Prior (✉)
Department of Business, Universitat Autònoma de Barcelona, Barcelona, Spain
e-mail: isabel.narbon@uab.es; diego.prior@uab.es

A. Khan et al. (eds.), *E-waste Recycling and Management*,
Environmental Chemistry for a Sustainable World 33,
https://doi.org/10.1007/978-3-030-14184-4_7

Keywords Cross-country comparison · Data Envelopment Analysis · Efficiency measurement · Environmental regulation · Environmental challenges · European Directives · European Union · E-waste · Management policy · Nonparametric methods · Waste electrical and electronic equipment

7.1 Introduction

During the last decades, sales of electrical and electronic equipment (henceforth, EEE) have increased exponentially worldwide (Pérez-Belis et al. 2015). This is partly due to technological innovations and new applications of EEE, which have become an essential part of human existence and day-to-day's life (Ylä-Mella et al. 2014). However, the replacement of obsolete equipment has accelerated because of the expansion of the innovation cycles, also making EEE a fast-growing source of waste. As a consequence, the management of the increasing volume of waste electrical and electronic equipment (henceforth, WEEE or E-waste) that EEE generates at its end-of-life has become a serious pollution problem (Kiddee et al. 2013), given its potential toxicity to environment and human health if disposal protocols are not properly managed (Townsend 2011). Under these circumstances, the effective and responsible management of WEEE has come to be a global major concern, being a priority target area to be regulated.

Accordingly, to address the environmental problems associated with the treatment and disposal of this particular form of hazardous waste, many countries have passed a specific legislation in order to ensure the proper management of WEEE (Pérez-Belis et al. 2015).[1] Specifically, they aim to preserve and protect the environment and the human health and to operate with prudency natural resources. In the European case, the framework for WEEE management is enacted in the Directive 2012/19/EU.[2] Its main objective is to contribute to (i) the prevention of WEEE; (ii) its reduction by encouraging the reuse, recycling, or recovery of such wastes; and (iii) the efficient use of resources and the retrieval of valuable secondary raw materials. To this end, it pursues the involvement of all operators associated with the life cycle of EEE, i.e., producers, distributors, consumers, and those operators directly involved in the collection and treatment of WEEE.

Therefore, with the purpose of fulfilling the objectives of the European Union's environment policy and monitor the compliance of countries, the Directive introduces ambitious collection targets. They are based on the amount of WEEE generated (*Article 7*) and the reuse, recycling, and recovery of WEEE separately collected after proper treatment (*Article 11*). Specifically, the Directive in its Annex V sets out specific minimum quantitative recovery targets for different categories of WEEE, evolving and increasing gradually over time as shown in Table 7.1. According to this, Member States should adapt their national policies on the management of

[1] See Ongondo et al. (2011) for a comprehensive review of waste electrical and electronic equipment (WEEE) management regulations in various countries.

[2] This Directive supplements the previous waste management legislation of the European Union, namely, Directive 2002/95/EC, Directive 2008/98/EC, and Directive 2009/125/EC.

Table 7.1 Minimum recovery targets by category of WEEE set out on the annex V of the Directive 2012/19/EU

Categories	From 13 August 2012 until 14 August 2015		From 15 August 2015 until 14 August 2018	
	Recovered	Recycled	Recovered	Recycled
Large household appliances Automatic dispensers	80%	75%	85%	80%
IT and telecommunications equipment Consumer equipment and photovoltaic panels	75%	65%	80%	70%
Small household appliances Lighting equipment Electrical and electronic tools Toys, leisure and sports equipment Medical devices Monitoring and control instruments	70%	50%	75%	55%
Gas discharge lamps	–	80%	–	80%

Notes: From 15 August 2018, a change in the categories of WEEE takes place and the minimum targets applicable by category referred to the new classification and not comparable

WEEE for compliance with the collection requirements and for the achievement of the recycling and recovery targets. Therefore, taking this idea as our starting point, the objective of this study is to analyze the performance of the European framework for WEEE management across different European Union countries, a subject yet to be examined in depth as we will see in the following paragraphs.

To date, since environmental concerns about WEEE first emerged, many researchers have investigated the situation of WEEE management from different points of view and contexts.[3] Indeed, we can find a large body of literature covering different areas such as consumer behavior toward WEEE (e.g., Saphores et al. 2006; Chi et al. 2014; Colesca et al. 2014; Wang et al. 2016; Borthakur and Govind 2017; Pérez-Belis et al. 2017), producers' positions toward legislative requirements (e.g., Stevels et al. 1999; Goosey 2004; Yu et al. 2006), the environmental and human health impacts of end-of-life WEEE (e.g., Wang and Guo 2006; Barba-Gutiérrez et al. 2008; Robinson 2009; Wäger et al. 2011; Kiddee et al. 2013), or the economic and environmental feasibility of recycling and reuse (e.g., Truttmann and Rechberger 2006; Gregory and Kirchain 2008; Kiatkittipong et al. 2008; Liu et al. 2009; Achillas et al. 2013; Cucchiella et al. 2015).

Moreover, other studies focused on the different management practices for treating WEEE (e.g., Townsend 2011; Kiddee et al. 2013; Shumon et al. 2014) or the legislation and regulation management systems in each country (e.g., Ongondo et al. 2011; Khetriwal et al. 2011; Zeng et al. 2013; Li et al. 2013) and its effective application (e.g., De Oliveira et al. 2012; Torretta et al. 2013; Popescu 2014; Ylä-Mella et al. 2014). However, to the best of our knowledge, the quantitative comparison of the effective management of WEEE of different countries according to the targets set by environmental regulations has not been measured before.

[3] See Pérez-Belis et al. (2015) for a comprehensive literature review on the main areas of research on WEEE, including studies published between 1992 and August 2014.

As regards the alternatives to measure management performance, methods from the perspective of productive efficiency have been widely applied in some fields of environmental performance. These techniques allow comparing the relative management performance of different decision-making units, providing helpful information for managers and policy-makers in order to design better managerial strategies and environmental policies.

In this context, some studies have addressed the issue of assessing performance through the concept of economic-ecological efficiency, commonly known as *eco-efficiency*. It refers to the ability of public and private organization to produce goods and services while incurring less impact on the environment and consuming fewer natural resources (e.g., Färe et al. 1989; Picazo-Tadeo and Prior 2009; Picazo-Tadeo et al. 2012; Sueyoshi and Goto 2011). In doing so, they consider environmental externalities as undesirable outputs (or bad outputs) in the production process. Moreover, other studies have focused on the performance measurement of several economic agents when managing different types of waste, notably including municipal solid waste (e.g., Marques and Simões 2009; Rogge and De Jaeger 2012) or wastewater (e.g., Abbott and Cohen 2009; Sala-Garrido et al. 2011). However, such an efficiency analysis has not been applied in previous literature to measure the management of WEEE or E-waste, on which we focus.

The aim of this study is to contribute to the literature on the management of WEEE by comparing the performance of different EU countries according to the targets set in the regulation of the Union's environmental policy on WEEE. To this end, we use the traditional nonparametric Data Envelopment Analysis (DEA) in order to measure technical efficiency for the first time in the literature. We use a sample of 30 EU countries for the year 2014, with the purpose of comparing their performance, ranking the countries, and identifying their level of inefficiency.

The results indicate that, as a general picture, European countries are highly efficient in the implementation of the WEEE. However, considering the performance of different waste categories, we observe significant differences in WEEE efficiency among countries. This means that, in the framework of the European Union, more effective control and regulation of the waste is required. In order to do so, as the multidimensional comparison among countries is not an easy task, nonparametric methods can be an effective tool for the European regulators.

The paper is organized as follows: Sect. 7.2 gives an overview of the methodology used to determine technical efficiency. Section 7.3 describes the data in detail. Section 7.4 presents and comments on the most relevant results. Finally, Sect. 7.5 summarizes the main conclusions.

7.2 Methodology

Productivity and efficiency are recurrent concepts in the literature of economics and management. On the one hand, productivity is an indicator that comes from engineering. In its most simple definition, productivity is calculated by dividing the physical units of production (outputs) by the physical units of factors consumed (inputs). So, the more the productivity, the more the production per unit of input consumed. The concept of efficiency is more demanding, as to be fully efficient, your level of productivity has to be the best possible. In other words, to do estimations of productive efficiency, you need to compare your own level of productivity with that coming from other producers. Once verified no other producer has the ability to operate with better productivity than you, then you can confirm you are fully efficient. This is the concept of productive efficiency (or technical efficiency); there are other possible definitions as, for instance, cost revenue or profit efficiency when we add to the physical units of outputs and inputs additional variables referring mainly to the output and input prices.

Paying attention to the existent seminal works, one can cite Koopmans (1951). Koopmans provided a well-founded definition of technical efficiency by indicating "a producer is technically efficient if an increase in any output requires a reduction in at least one other output or an increase in at least one input, and if a reduction in any input requires an increase in at least one other input or a reduction in at least one output" (Koopmans (1951), page 60).

There are other formal definitions of technical efficiency. For instance, it is worth citing Debreu (1951) and Farrell (1957) who introduced a less demanding indicator of technical efficiency that refers to the maximum radial (proportional) reduction in all inputs that continues providing a determined level of output. This means that a score of unity serves to identify the existence of technical efficiency. When this is not the case, a score less than unity serves to exhibit the presence of technical inefficiency. This Debreu–Farrell definition is based on the inputs contraction (meaning, it is an *input-oriented* indicator). Reversing, the terms, we can also define an *output-oriented* indicator by accepting an expansion of the output variables by keeping constant the inputs.

So, different than productivity, efficiency indicators determine the existent distance between a determined level of outputs and inputs and the specific level of outputs and inputs that are required to define the best possible combination that serves to configure the best practice frontier of the existent technology. As indicated, the input-oriented indicator exhibits the necessary input reduction to become efficient, keeping constant the outputs. In contrast, the output-oriented indicator signals the required output expansion but keeping constant the inputs. There are other extensions, the so-called directional distance functions, that combine changes in inputs and outputs, but their formal definitions are more complex, and we will not use them in our empirical work.

Let's start with the definition process of the indicator of efficiency (or eco-efficiency) we are going to follow in the empirical work. To do this, we follow Farrell who defined his indicator of technical efficiency by postulating a convex technology of the production side. This initial definition has been expanded in order to provide more possible methodological choices:

1. Deterministic frontier analysis by using parametric production functions (DFA). The seminal work on this direction is Aigner and Chu (1968).

2. Stochastic frontier analysis (SFA) (Aigner et al. 1977; (Meeusen and Van den Broek 1977). This method estimates efficiency by using stochastic production functions.
3. Data Envelopment Analysis (DEA), originally proposed by Charnes et al. (1978) which measures efficiency relative to a deterministic nonparametric frontier by using mathematical optimization methods.

Each of these methodologies has its advantages and disadvantages, as can be seen in the excellent review provided in Lovell (1996). Particularly, we are going to use DEA estimation method because it requires the minimal level of assumptions regarding the technology.

Now we define the mathematical models required to do the estimations of the scores of efficiency. Assume that we have K observations. A specific unit k, ($k = 1, \ldots, K$) produces a given amount of output y_{km} ($m = 1, \ldots, M$) by consuming x_{kn} ($n = 1, \ldots, N$) inputs. We also assume to know the matrix Y of the M outputs for the K units (this means that the outputs matrix has KxM dimensions). We also assume the knowledge of the matriz X of the N observed inputs corresponding to the K units (this means that the inputs is defined with dimension KxN). Having this information, we define the input vectors (x_n) that are consumed in the production of the output vector (y_m); we also assume to know the technology that allows to transform the inputs into the outputs. As mentioned, we take the output-oriented version of the efficiency estimation. This technology starts from the definition of the output set. Shephard (1970) has proven that a linear technology exhibiting the usual properties (regularity, monotonicity, convexity, and variable returns to scale) can be summarized by the output set:

$$F(x) = \{y : (x,y) \text{ is feasible}\} \tag{7.1}$$

The output set includes all possible input and output sets, meaning inefficient and efficient points. If we take a more demanding position, the isoquant offers the Debreu–Farrell notion of efficiency:

$$\text{Isoq } F(x) = \{y : y \in F(x), zy \notin F(x), z \in (1,+\infty)\} \tag{7.2}$$

where z is the intensity vector and $F(x_k)$ includes the inputs required to produce the output vector y_k. From the isoquant, now it is possible to operationalize the Debreu–Farrell output-oriented measure of technical efficiency:

$$DF_o(x,y) = \max\{\theta : \theta.y \in F(x)\} \tag{7.3}$$

Knowing θ it is easy to determine the potential or optimal level of output (the one that projects the observed unit k on the efficient frontier):

$$\theta.y_k, \text{ with } \theta \geq 1 \tag{7.4}$$

If $\theta = 1$, then y_k performs on the efficient frontier, while if $\theta > 1$, then it is needed an increase in the level of output output vector y_k to find the efficiency frontier. To quantify the output efficient score, θ, we could follow two ways: operate with frontier production functions (parametric approach) or operate without imposing any functional form in particular (nonparametric DEA models). We apply here the nonparametric approach. The linear programming problem used to calculate the efficient frontier for the output-oriented efficiency coefficient known as Data Envelopment Analysis (DEA) in variable returns to scale version (VRS) is the following:

$$\begin{aligned} DF_o\left(x^o, y^o\right) = \; & \max_{\theta, z_k} \theta \\ \text{subject to} \quad & \sum_{K}^{k=1} z_k y_{km} \geq \theta y_m^o \qquad m = 1, \ldots M, \\ & \sum_{K}^{k=1} z_k x_{kn} \leq x_n^o \qquad n = 1, \ldots N, \\ & \sum_{K}^{k=1} z_k = 1, \\ & z_k \geq 0, \qquad k = 1, \ldots K. \end{aligned} \tag{7.5}$$

Note that the restriction from 7.5: $\sum_{K}^{k=1} z_k = 1$ corresponds to the DEA–VRS program, meaning the technology exhibits variable returns to scale.

In Fig. 7.1 we can observe the inefficiency level of unit B_k.This inefficiency is denoted by the vertical distance separating the best practice output frontier from the observed output level.

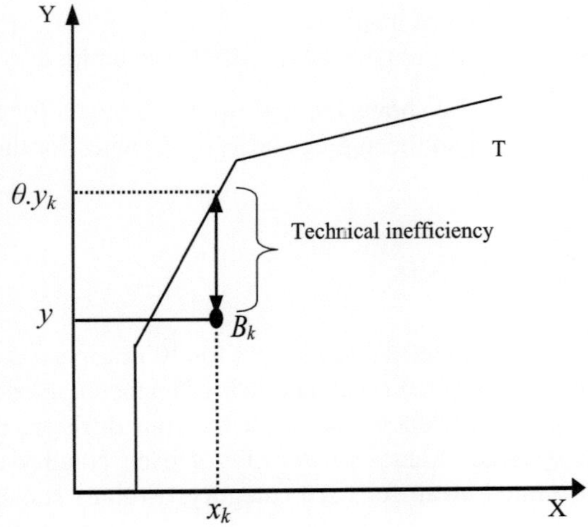

Fig. 7.1 Technical inefficiency with variable returns to scale technology (This figure represents the variable returns to scale output-oriented Data Envelopment Analysis (DEA) model. Technical inefficiency level of unit B_k is denoted by the vertical distance separating the best practice output frontier from the observed output level)

7.3 Sample, Data, and Variables

We carry out the analysis for a sample of European Union Member State and European Economic Area (EEA) and the European Free Trade Association (EFTA) countries. The information on inputs and outputs comes from the Eurostat, which is the statistical office of the European Union. Specifically, we use data on waste electrical and electronic equipment (WEEE) that includes information about the compliance of countries with the minimum quantitative targets applicable for reuse, recycling, and recovery of WEEE, after proper treatment. It was collected on the basis of Directive 2012/19/EU of the European Parliament and of the Council of 4 July 2012. This data has been published annually since 2005; however, until the reference year 2013, some countries, such as Austria, Bulgaria, Czech Republic, Malta, the Netherlands, and the United Kingdom, presented data with anomalies or changes in methodologies to measure some particular categories. As a consequence, data before that year was not comparable for all countries.

Our input measure represents the tons of WEEE which has been collected and properly treated by each country (X_1). As for the output variables, we have chosen two different measures, namely, the tons of WEEE recovered (Y_1) and the tons of WEEE reused or recycled (Y_2). Moreover, we differentiate four different groups of WEEE, defined according to the categories listed in the Directive 2012/19/EU. It sets out specific minimum recovery targets for four different categories of WEEE for the period from 13 August 2012 until 14 August 2015 (see Table 7.1 for further details). Specifically, the groups are defined as follows:

Category A contains (i) large household appliances and (ii) automatic dispensers.
Category B contains (i) IT and telecommunications equipment, and (ii) consumer equipment and photovoltaic panels.
Category C contains (i) small household appliances, (ii) lighting equipment (with the exception of gas discharge lamps), (iii) electrical and electronic tools, (iv) toys, (v) leisure and sports equipment, and (vi) medical devices and monitoring and control instruments.
Category D contains (i) gas discharge lamps.

Table 7.2 shows the descriptive statistics for inputs and outputs in each WEEE category and the aggregated WEEE values for the year 2014.

7.4 Results

We estimate efficiency scores for 30 European countries for the year 2014 obtained by applying the nonparametric efficiency approach previously described. Table 7.3 reports efficiency results for the four different categories of WEEE and the total aggregated values of WEEE for each country under analysis. It also shows the trimmed mean for each category, a robust statistical measure of central tendency

Table 7.2 Descriptive statistics for inputs and outputs

Category[a]	Variables[b]	Mean	Median	S.d.
A	Collected (X_1)	58445.80	29531.50	82368.93
	Recovery (Y_1)	52350.87	25735.00	74860.14
	Recycling reuse (Y_2)	47933.67	24177.00	66999.95
B	Collected (X_1)	42064.73	16464.00	65733.78
	Recovery (Y_1)	37159.47	14399.50	60037.82
	Recycling reuse (Y_2)	34197.60	13562.00	54456.01
C	Collected (X_1)	18031.70	6437.00	33320.08
	Recovery (Y_1)	16129.53	5670.50	31755.08
	Recycling reuse (Y_2)	14564.90	5604.50	27463.60
D	Collected (X_1)	1258.57	664.50	1732.77
	Recycling reuse (Y_2)	1117.83	588.50	1558.03
Total	Collected (X_1)	119800.77	55180.00	176218.35
	Recovery (Y_1)	106765.80	48305.00	161418.90
	Recycling reuse (Y_2)	97812.87	45752.00	144963.98

[a]Categories are defined according to the Annex V of Directive 2012/19/EU of the European Parliament and of the Council of 4 July 2012 which sets out the minimum recovery targets for by category of WEEE
[b]In tons

that offers an estimation of the mean after discarding both tails of the distribution to avoid outliers distortion. Moreover, we also provide violin plots for further interpretation of results. They include all features of the distribution and offer more thorough information on how the categories of WEEE behave. Specifically, Fig. 7.2 shows the violin plots for category A and the total aggregated amount of WEEE.[4]

The results show that the mean technical efficiency score for all WEEE categories and all European countries is 1.063. It suggests that they are properly managing WEEE but the total amount of EEE collected and treated for recovering and recycling could still be improved by 6.3%. However, if we compare the results yielded by different countries, in general, they show several discrepancies.

Focusing on the best performing countries, we observe that only Germany presents an efficiency score of 1 for all the categories, i.e., it is always efficient. Moreover, both Croatia and Malta are efficient for all the WEEE categories except for category D, in which they are highly inefficient. Nevertheless, despite its very poor management in this particular category, both countries are considered as quite efficient units as a whole. This is due to this type of waste presents the minor weight in terms of tons from the total aggregated set of WEEE. Finally, some countries such as the Netherlands and Finland present some room for improvement in most of the categories, but they are efficient when regarding the total WEEE. In contrast, Iceland and Lithuania are the worst performing countries given their high inefficiency

[4]For visual simplicity, we only plot the total aggregated amount of WEEE together with category A, which is the most important category of WEEE in terms of tones; however, individual plots for the rest of categories are available upon request.

Table 7.3 Efficiency results for the different categories of waste electrical and electronic equipment (WEEE)

Country	Category A	Category B	Category C	Category D	TOTAL
Austria (AT)	1.029	1.075	1.044	1.140	1.048
Belgium (BE)	1.124	1.106	1.184	1.112	1.115
Bulgaria (BG)	1.007	1.238	1.131	1.259	1.051
Cyprus (CY)	1.137	1.358	1.066	1.000	1.159
Czech Republic (CZ)	1.036	1.079	1.153	1.152	1.031
Germany (DE)	1.000	1.000	1.000	1.000	1.000
Denmark (DK)	1.172	1.000	1.011	1.085	1.046
Estonia (EE)	1.268	1.137	1.067	1.256	1.122
Greece (EL)	1.094	1.032	1.066	1.434	1.059
Spain (ES)	1.139	1.064	1.198	1.350	1.108
Finland (FI)	1.028	1.019	1.001	1.118	1.000
France (FR)	1.000	1.042	1.026	1.091	1.043
Croatia (HR)	1.000	1.000	1.000	1.487	1.000
Hungary (HU)	1.041	1.039	1.000	1.221	1.017
Ireland (IE)	1.069	1.092	1.072	1.177	1.063
Iceland (IS)[b]	1.000	2.030	9.887	–	1.308
Italy (IT)	1.000	1.069	1.100	1.080	1.028
Lithuania (LT)	1.159	1.326	1.190	1.321	1.196
Luxembourg (LU)	1.126	1.206	1.152	1.143	1.087
Latvia (LV)	1.154	1.266	1.089	1.209	1.093
Malta (MT)[b]	1.000	1.000	1.000	–	1.000
Netherlands (NL)	1.000	1.001	1.009	1.084	1.000
Norway (NO)	1.014	1.081	1.030	1.131	1.029
Poland (PL)	1.010	1.743	1.067	1.135	1.160
Portugal (PT)	1.113	1.201	1.113	1.325	1.122
Romania (RO)	1.000	1.137	1.081	1.066	1.033
Sweden (SE)	1.017	1.004	1.112	1.000	1.017
Slovenia (SI)	1.047	1.226	1.076	1.000	1.066
Slovakia (SK)	1.000	1.073	1.028	1.164	1.019
United Kingdom (UK)	1.000	1.054	1.003	1.090	1.047
Trimmed mean efficiency scores[a]	1.054	1.131	1.070	1.159	1.063

[a]A truncated mean or trimmed mean is a robust statistical measure of central tendency, which involves the calculation of the mean after discarding both tails of the distribution
[b]Both Iceland and Malta do not collect waste for category D

level in most categories. Moreover, Poland, Cyprus, and Estonia are also highly inefficient given their poor management in categories A and B, which are the most important ones (in terms of tons) for the total aggregated set of WEEE.

Otherwise, regarding the different types of WEEE, we observe that some categories are better managed than others. Category A, which corresponds to large household appliances and automatic dispensers, is the one that presents the best average efficiency levels. Indeed, we find many countries which are totally efficient in this

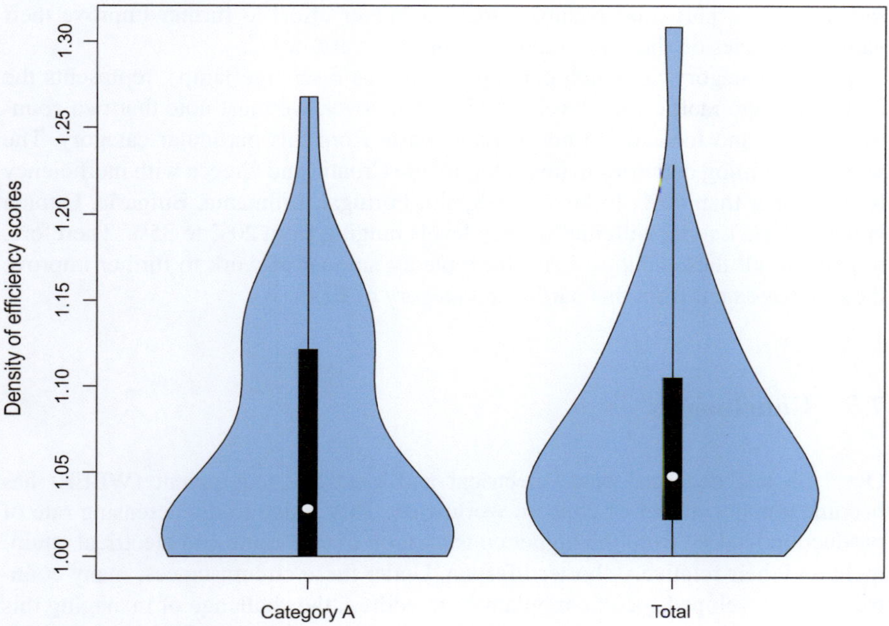

Fig. 7.2 Violin plots for category A and total WEEE
This figure presents violin plots for both the total aggregated amount of waste electrical and electronic equipment (WEEE) and category A, which corresponds to large household appliances and automatic dispensers

category, such as Germany, Croatia, Malta, the Netherlands, Slovakia, Italy, Romania, France, the United Kingdom, and Iceland. The good performance in this particular category partly explains the high efficiency levels for the total WEEE categories. Figure 7.2 supports descriptive analysis. The category A's figure shows a bimodal structure in which the greater probability mass is concentrated around the unity, and it shows more quantity of efficient units than the figure for the total aggregated amount of WEEE.

Category B represents the most important category after category A. It contains telecommunications equipment and consumer equipment and photovoltaic panels. The average efficiency score for this category B is 1.131. This high inefficiency level could be explained by the poor performance of some countries, including Iceland, Poland, Cyprus, Lithuania, and Latvia, which could improve their inefficiency levels more than 25%. Only Germany, Croatia, Malta, and Denmark are efficient in this category. Moreover, from category C we must emphasize the extremely poor performance from Iceland. Although for the rest of countries the inefficiency levels are not so high, countries such as Spain, Lithuania, Belgium, and Luxembourg present inefficiency scores above 15%. Therefore, it would be necessary that these particular countries make a bigger effort to further improve their national policies on the waste management from category C.

Finally, category D, which corresponds to gas discharge lamps, represents the higher average inefficiency levels (1.159). Moreover, we must note that two countries, Malta and Iceland, do not manage waste from this particular category. The worst performing countries in this category are Croatia and Greece with inefficiency levels higher than 40%, followed by Spain, Portugal, Lithuania, Bulgaria, Estonia Hungary, and Latvia, with inefficiency levels ranging from 20% to 35%. Therefore, in general, all these countries still have plenty amount of work to further improve their management from this particular category of EEE.

7.5 Conclusions

Over the last decades, waste electrical and electronic equipment (WEEE) has become a major matter of concern worldwide. This is due to its increasing rate of production leading from the higher consumption of electronic and electrical equipment and their relatively shorter lifespan. Under these circumstances, many countries have developed specific regulations to address the challenge of managing this environmental issues. In the European case, the framework for WEEE management was enacted in the Directive 2012/19/EU. Its main objective is to contribute to (i) the prevention of WEEE; (ii) its reduction by encouraging the reuse, recycling, or recovery of WEEE; and (iii) the efficient use of resources by reusing valuable raw materials from the obsolete equipment. According to this, it introduces targets for the reuse and recycling and recovery of WEEE separately collected after proper treatment, which European countries should include to their national policies.

The purpose of this study was to contribute to the literature on waste management of EEE by comparing the performance of different European countries according to the implementation of the targets set by the European Union Directive on WEEE. We evaluate it from the perspective of productive efficiency. To the best of our knowledge, this article is the first attempt to assess this dimension of European eco-efficiency, although there are more works to be done in the near future.

It can be concluded that, in general, European countries have efficiently succeeded in the implementation of the WEEE Directive and management of the recycling and recovery of WEEE. However, regarding the performance of the different countries and waste categories, there are large differences in their WEEE management. This means that there is room to increase the performance, as some countries' WEEE management systems need improvements to meet the European Directives requirements. We have to recognize that, in some countries, the WEEE management system was adapted to the European legislative changes as much as possible, but its impact is, so far, limited as the targets are not yet totally accomplished. Summing up, bigger efforts made by public authorities and responsible operators are due in the near future.

With the aims to help the plans to improve the level of eco-efficiency in the EEE, our analysis has identified some potential areas of improvement. Specifically, more efforts are needed to achieve higher efficiency levels from waste categories C and D, which mainly contains small equipment, lights, electrical and electronic tools, and medical devices.

The implications that can be extracted from our research are obvious, as some countries have to face a real challenge in the treatment of WEEE. This challenge could be achieved by better adapting national policies to the European legislative requirements through more convenient collection and treatment for this kind of EEE; additional containers in public places, at retailers; public awareness campaigns supporting consumption and collection; encouragement for the design and production of EEE which take into full account its repair, reuse, disassembly, and recycling; better control from public institutions regarding their recycling and recovery; and stronger legislation regarding the obligations of producers and distributors, among others.

From the empirical research point of view, extensions of this investigation are expected. Among them, in our opinion two of them are relevant both from the theoretical and the empirical perspectives: (a) the consideration of a more complex technology by considering the existence of *bad* or *undesirable* outputs, i.e., the tons of material collected but neither recovered nor recycled, and (b) the extension of the mathematical models by considering the process of adjustment and target setting in the efficiency estimations. We hope to spend effort and attention on these extensions in the near future.

References

Abbott M, Cohen B (2009) Productivity and efficiency in the water industry. Util Policy 17(3–4):233–244. https://doi.org/10.1016/j.jup.2009.05.001

Achillas C, Aidonis D, Vlachokostas C, Karagiannidis A, Moussiopoulos N, Loulos V (2013) Depth of manual dismantling analysis: a cost–benefit approach. Waste Manag 33(4):948–956. https://doi.org/10.1016/j.wasman.2012.12.024

Aigner DJ, Chu S-F (1968) On estimating the industry production function. Am Econ Rev 58(4):826–839

Aigner D, Lovell CK, Schmidt P (1977) Formulation and estimation of stochastic frontier production function models. J Econ 6(1):21–37. https://doi.org/10.1016/0304-4076(77)90052-5

Barba-Gutiérrez Y, Adenso-Diaz B, Hopp M (2008) An analysis of some environmental consequences of European electrical and electronic waste regulation. Resour Conserv Recycl 52(3):481–495. https://doi.org/10.1016/j.resconrec.2007.06.002

Borthakur A, Govind M (2017) Emerging trends in consumers' E-waste disposal behaviour and awareness: a worldwide overview with special focus on India. Resour Conserv Recycl 117:102–113. https://doi.org/10.1016/j.resconrec.2016.11.011

Charnes A, Cooper WW, Rhodes E (1978) Measuring the efficiency of decision making units. Eur J Oper Res 2(6):429–444. https://doi.org/10.1016/0377-2217(78)90138-8

Chi X, Wang MY, Reuter MA (2014) E-waste collection channels and household recycling behaviors in Taizhou of China. J Clean Prod 80:87–95. https://doi.org/10.1016/j.jclepro.2014.05.056

Colesca S, Ciocoiu C, Popescu M (2014) Determinants of WEEE recycling behaviour in Romania: a fuzzy approach. Int J Environ Res 8(2):353–366. https://doi.org/10.22059/IJER.2014.726

Cucchiella F, D'Adamo I, Koh SL, Rosa P (2015) Recycling of WEEEs: an economic assessment of present and future e-waste streams. Renew Sust Energ Rev 51:263–272. https://doi.org/10.1016/j.rser.2015.06.010

De Oliveira CR, Bernardes AM, Gerbase AE (2012) Collection and recycling of electronic scrap: a worldwide overview and comparison with the Brazilian situation. Waste Manag 32(8):1592–1610. https://doi.org/10.1016/j.wasman.2012.04.003

Debreu G (1951) The coefficient of resource utilization. Econometrica 19(3):273–292. https://doi.org/10.2307/1906814

Färe R, Grosskopf S, Logan J (1985) The relative performance of publicly-owned and privately-owned electric utilities. J Public Econ 26(1):89–106. https://doi.org/10.1016/0047-2727(85)90040-4

Färe R, Grosskopf S, Lovell CK, Pasurka C (1989) Multilateral productivity comparisons when some outputs are undesirable: a nonparametric approach. Rev Econ Stat 71:90–98. https://doi.org/10.2307/1928055

Farrell M (1957) The measurement of productive efficiency. J R Stat Soc 120(III):253–281. https://doi.org/10.2307/2343100

Goosey M (2004) End-of-life electronics legislation–an industry perspective. Circuit World 30(2):41–45. https://doi.org/10.1108/03056120410512235

Gregory JR, Kirchain RE (2008) A framework for evaluating the economic performance of recycling systems: a case study of North American electronics recycling systems. Environ Sci Technol 42(18):6800–6808. https://doi.org/10.1021/es702666v

Khetriwal DS, Widmer R, Kuehr R, Huisman J (2011) One WEEE, many species: lessons from the European experience. Waste Manag Res 29(9):954–962. https://doi.org/10.1177/0734242X11413327

Kiatkittipong W, Wongsuchoto P, Meevasana K, Pavasant P (2008) When to buy new electrical/electronic products? J Clean Prod 16(13):1339–1345. https://doi.org/10.1016/j.jclepro.2007.06.019

Kiddee P, Naidu R, Wong MH (2013) Electronic waste management approaches: an overview. Waste Manag 33(5):1237–1250. https://doi.org/10.1016/j.wasman.2013.01.006

Koopmans TC (1951) Analysis of production as an efficient combination of activities. Anal Prod Alloc 173(3):33–97. https://doi.org/10.2307/2226909

Li J, Liu L, Zhao N, Yu K, Zheng L et al (2013) Regional or global WEEE recycling. Where to go? Waste Manag 33(4):923–934. https://doi.org/10.1016/j.wasman.2012.11.011

Liu X, Tanaka M, Matsui Y (2009) Economic evaluation of optional recycling processes for waste electronic home appliances. J Clean Prod 17(1):53–60. https://doi.org/10.1016/j.jclepro.2008.03.005

Lovell CK (1996) Applying efficiency measurement techniques to the measurement of productivity change. J Prod Anal 7(2–3):329–340. https://doi.org/10.1007/BF00157047

Marques RC, Simões P (2009) Incentive regulation and performance measurement of the Portuguese solid waste management services. Waste Manag Res 27(2):188–196. https://doi.org/10.1177/0734242X08095025

Meeusen W, Van den Broek J (1977) Efficiency estimation from Cobb-Douglas production function with composed error. Int Econ Rev 8:435–444. https://doi.org/10.2307/2525757

Ongondo FO, Williams ID, Cherrett TJ (2011) How are WEEE doing? A global review of the management of electrical and electronic wastes. Waste Manag 31(4):714–730. https://doi.org/10.1016/j.wasman.2010.10.023

Pérez-Belis V, Bovea M, Ibáñez-Forés V (2015) An in-depth literature review of the waste electrical and electronic equipment context: trends and evolution. Waste Manag Res 33(1):3–29. https://doi.org/10.1177/0734242X14557382

Pérez-Belis V, Braulio-Gonzalo M, Juan P, Bovea MD (2017) Consumer attitude towards the repair and the second-hand purchase of small household electrical and electronic equipment. A Spanish case study. J Clean Prod 158:261–275. https://doi.org/10.1016/j.jclepro.2017.04.143

Picazo-Tadeo AJ, Prior D (2009) Environmental externalities and efficiency measurement. J Environ Manag 90(11):3332–3339. https://doi.org/10.1016/j.jenvman.2009.05.015

Picazo-Tadeo AJ, Beltrán-Esteve M, Gómez-Limón JA (2012) Assessing eco- efficiency with directional distance functions. Eur J Oper Res 220(3):798–809. https://doi.org/10.1016/j.ejor.2012.02.025

Popescu M-L (2014) Waste electrical and electronic equipment management in Romania: harmonizing national environmental law with European legislation. Administratie si Management Public 188(22):65–72. https://doi.org/10.1016/j.sbspro.2015.03.389

Robinson BH (2009) E-waste: an assessment of global production and environmental impacts. Sci Total Environ 408(2):183–191. https://doi.org/10.1016/j.scitotenv.2009.09.044

Rogge N, De Jaeger S (2012) Evaluating the efficiency of municipalities in collecting and processing municipal solid waste: a shared input DEA-model. Waste Manag 32(10):1968–1978. https://doi.org/10.1016/j.wasman.2012.05.021

Sala-Garrido R, Molinos-Senante M, Hernández-Sancho F (2011) Comparing the efficiency of wastewater treatment technologies through a DEA metafrontier model. Chem Eng J 173(3):766–772. https://doi.org/10.1016/j.cej.2011.08.047

Saphores J-DM, Nixon H, Ogunseitan OA, Shapiro AA (2006) Household willingness to recycle electronic waste: an application to California. Environ Behav 38(2):183–208. https://doi.org/10.1177/0013916505279045

Shephard R (1970) Theory of cost and production function. Princeton University Press. https://doi.org/10.2307/2230285

Shumon MRH, Ahmed S, Islam MT (2014) Electronic waste: present status and future perspectives of sustainable management practices in Malaysia. Environ Earth Sci 72(7):2239–2249. https://doi.org/10.1007/s12665-014-3129-5

Stevels A, Ram A, Deckers E (1999) Take-back of discarded consumer electronic products from the perspective of the producer: conditions for success. J Clean Prod 7(5):383–389. https://doi.org/10.1016/S0959-6526(99)00152-3

Sueyoshi T, Goto M (2011) Measurement of returns to scale and damages to scale for DEA-based operational and environmental assessment: how to manage desirable (good) and undesirable (bad) outputs? Eur J Oper Res 211(1):76–89. https://doi.org/10.1016/j.ejor.2010.11.013

Torretta V, Ragazzi M, Istrate IA, Rada EC (2013) Management of waste electrical and electronic equipment in two EU countries: a comparison. Waste Manag 33(1):117–122. https://doi.org/10.1016/j.wasman.2012.07.029

Townsend TG (2011) Environmental issues and management strategies for waste electronic and electrical equipment. J Air Waste Manag Assoc 61(6):587–610. https://doi.org/10.3155/1047-3289.61.6.587

Truttmann N, Rechberger H (2006) Contribution to resource conservation by reuse of electrical and electronic household appliances. Resour Conserv Recycl 48(3):249–262. https://doi.org/10.1016/j.resconrec.2006.02.003

Wäger P, Hischier R, Eugster M (2011) Environmental impacts of the Swiss collection and recovery systems for waste electrical and electronic equipment (WEEE): a follow-up. Sci Total Environ 409(10):1746–1756. https://doi.org/10.1016/j.scitotenv.2011.01.050

Wang J, Guo X (2006) Impact of electronic wastes recycling on environmental quality. Biomed Environ Sci 19(2):137. PMID: 16827186

Wang Z, Guo D, Wang X (2016) Determinants of residents' e-waste recycling behaviour intentions: evidence from China. J Clean Prod 137:850–860. https://doi.org/10.1016/j.jclepro.2016.07.155

Ylä-Mella J, Poikela K, Lehtinen U, Keiski RL, Pongrácz E (2014) Implementation of waste electrical and electronic equipment directive in Finland: evaluation of the collection network and challenges of the effective WEEE management. Resour Conserv Recycl 86:38–46. https://doi.org/10.1016/j.resconrec.2014.02.001

Yu J, Welford R, Hills P (2006) Industry responses to EU WEEE and ROHS directives: perspectives from China. Corp Soc Responsib Environ Manag 13(5):286–299. https://doi.org/10.1002/csr.131

Zeng X, Li J, Stevels A, Liu L (2013) Perspective of electronic waste management in China based on a legislation comparison between China and the EU. J Clean Prod 51:80–87. https://doi.org/10.1016/j.jclepro.2012.09.030

Chapter 8
E-Waste Management from Macroscopic to Microscopic Scale

Chukwudi O. Onwosi, Victor C. Igbokwe, Tochukwu N. Nwagu, Joyce N. Odimba, and Charles O. Nwuche

Contents

C. O. Onwosi (✉) · V. C. Igbokwe · J. N. Odimba
Department of Microbiology, Faculty of Biological Sciences, University of Nigeria, Nsukka, Enugu State, Nigeria

Bioconversion and Renewable Energy Research Unit, University of Nigeria, Nsukka, Enugu State, Nigeria
e-mail: chukwudi.onwosi@unn.edu.ng

T. N. Nwagu · C. O. Nwuche
Department of Microbiology, Faculty of Biological Sciences, University of Nigeria, Nsukka, Enugu State, Nigeria

A. Khan et al. (eds.), *E-waste Recycling and Management*, Environmental Chemistry for a Sustainable World 33,
https://doi.org/10.1007/978-3-030-14184-4_8

Abstract Increased demand for electrical and electronic equipment as well as a reduction in the end of life of most electrical products has led to the generation of large amount of E-waste. These wastes contain both beneficial and hazardous components. Therefore, there should be proper management of E-waste in order to protect man and the environment. In this review, we addressed the various categories deployed towards effective E-waste management such as collection and disposal of dangerous portions and recovery of precious metals and energy. The benefits, challenges and future of E-waste management were also highlighted.

Keywords E-waste · Recycling · A printed wiring board · Precious metals · Material flow analysis · Life cycle assessment · Metallurgy · Heavy metals · Plastic · Extended producer responsibility

8.1 Introduction

Implications of Uncontrolled Management of Waste Electronics

A lot of E-waste components are generated in many countries mostly in low- and middle-income households, and these E-wastes are often disposed into wild landfills. In some of these countries, including Nigeria, Ghana, India, Pakistan, China and the Philippines, E-wastes are dismantled using simple equipment to retrieve vital metals by unexpected and untrained persons without much consideration for environmental and public health (Ikhlayel 2018). In these countries, E-waste salvaging predominantly involves processes such as burning of wires in open spaces to retrieve copper from the interior while confiscating the plastic part, also the extraction of precious metals such as silver, gold and platinum from printed wiring boards.

Many reports in the literature have outlined the components of E-wastes which are known toxicants that have deleterious effects on human health. These toxicants including heavy metals and polychlorinated biphenyls (PCBs) have been shown to be rampant in sediments, air and aquatic biotas in regions where E-wastes abound. The negative effects of these toxicants on organisms range from acute effects, endocrine disruption, reproductive dysfunction and cancer (Zhang et al. 2014).

The management of electronic and electrical wastes (E-waste) in various regions of the world has received tremendous attention recently owing to the already established negative eco-impact(s) of these very hazardous waste materials. Several studies done in relation to E-waste management have shown that E-waste(s) management seems to be region-specific as many countries have implemented varied policies to check the rampaging of these wastes materials in that region (Townsend 2011).

The transboundary movement of E-waste from developed countries to developing countries has been a major issue of concern for a long time (Townsend 2011). There is still a great deal of work as regards E-waste management in developing countries. However, the opposite can be said in the case of developed nations

(Ikhlayel 2018). There are still arguments that E-waste generation in some developing countries does not call for urgent attention because of the lesser quantity and longer half-life of electronic goods in those countries due to financial constraints, on both local community and national scales (Kiddee et al. 2013).

In some developed countries, E-wastes have been treated or disposed of by landfill or by informal incineration as in the case of some developing countries. However, technological innovations and recycling methods have been developed for regional and global E-waste management (Zeng et al. 2015). In order to avert public debate, the cost of disposing of E-waste must be balanced against the environmental benefit (Zhang et al. 2014).

Different options for managing electronic-wastes ranging from the microscopic to the macroscopic outlook are discussed. However, it is necessary to note that the method governing regulatory structure in a region and the already established management options in a region are major factors that influence the choice of option for managing E-waste.

8.1.1 Macroscopic Management of E-Waste

8.1.1.1 The Role of Government in E-Waste Management

Opportunities obtainable for handling E-wastes depend greatly on location and existing governing regulatory structure (Townsend 2011). Different legislations (e.g. the Basel Convention) with regard to E-waste have been enacted to check the wider issues concerning E-waste management.

Compared to the more advanced countries, the state of affairs regarding E-waste is even more of a threat in the developing world. In order to curb this menace, most of the developed world and several countries in the developing world have endorsed legislation to check unlawful trafficking and unlicensed recycling of E-waste. These legislations appeal the extended producer responsibility concept based on life cycle considerations in the hope that it will provide inhibition as well as a remedy (Premalatha et al. 2014).

Laws and policies regarding the appropriate management of E-wastes continue to advance with time (Ramesh Babu et al. 2007). However, such policy designs should be innovative and should be strictly enforced in order to exert its efficacy (Lu et al. 2015). In the aspect of promulgating and then implementing E-waste regulations, the European Union and some Asian countries have been described as frontiers. Meanwhile, the Swiss are ascribed with founding the first comprehensive E-waste management system, covering collection to disposal (Sthiannopkao and Wong 2013).

Laws and guidelines have been enacted by the European Union to restrict the use of materials in electrical and electronic equipment which are detrimental to human health and environmental safety (Directive 2002/95/EC, the RoHs Directive). Guidelines were also stipulated on the promotion of E-waste recycling and collection processes (Directive 2002/96/EC, the WEEE Directive). These guidelines also include steps for the creation of 'collection scheme' for users of 'electrical and electronic' gadgets to deposit their wastes without payment. This is targeted towards increasing E-waste collection, recycling and/or reuse (Zhang et al. 2015).

Upon the end of life (EoL), the strategy used in managing E-wastes depends on the following:

- Applicable regulations and policies already in place (e.g. disposal bans, recycling bans)
- Existing infrastructure for handling such materials (e.g. availability of take-back centres, collection opportunities)
- Consumers' understanding of and attitudes towards these programs, policies and opportunities (Townsend 2011)

Apart from government interference and legislation, a lot of non-governmental bodies and individual groups have been involved in E-waste clean up; however, in most parts of the world, much success has not been attained (Premalatha et al. 2014).

Funding researches on E-waste management should be encouraged and adequately sponsored by private and government agencies. This will encourage more researchers to focus on this area, leading to novel ideas and technologies for effective E-waste collection and handling systems (Lu et al. 2015).

To achieve an efficient E-waste handling system, there is the need for active participation of all sectors involved, not just the manufacturers, distributors/suppliers and end-users but also those who collect, recycle or reuse waste items. This is to ensure that adequate and economical protocols are applied to equipment management and subsequent E-waste reuse and disposal (Tansel 2017).

8.1.1.2 The Role of Consumers in E-Waste Management

The awareness and behaviour of consumers towards the proper management of E-wastes are two key factors in any effective E-waste management strategy (Borthakur and Govind 2017). The socioeconomic profile of consumers, such as gender, age, income and education level, plays key roles in the behaviour or willingness of the general public to participate in managing E-wastes (Yin et al. 2014). Consumers of e-products also have a role to play in ameliorating the burden of E-waste in the environment. Consumers can minimize E-waste accumulation by:

- Using only necessary e-gadgets
- Efficient use of electrical/electronic equipment in order to extend the shelf life
- Purchasing items with very minor or zero E-wastes
- Avoiding addiction and dependency on electrical/electronic gadgets
- Being informed on the long-term negative impact of E-waste on the environment
- Pro-actively taking measures to achieve and maintain zero E-waste accumulation

Heightened public awareness concerning wastes such as E-wastes may create a possibility that consumers would be willing to pay some costs associated with recycling these waste materials. Payments could be in the form of prepaid deposit or incorporated into the product fee prior to the purchase of the equipment (Yin et al. 2014). In order to encourage public participation in the management of E-wastes, training sections and workshops should be organized, developed and publicized both in the form of television programs and newsletters, in order to further enlighten the public on steps to manage and contain E-waste. These activities can serve as a common playground for stakeholders to reinforce their mutual understanding, trust and respect, which will become a solid groundwork for a further partnership in the field of E-waste management (Lu et al. 2015).

During these programs, the awareness of the high environmental and health risks associated with improper handling of these wastes should be created and adequate remedies propagated for future purposes.

8.1.1.3 Extended Producer Responsibility

Extended producer responsibility (EPR) can be defined as an environmental protection strategy that makes the producer of a product responsible for the entire life cycle of the product and especially for the take-back, recycling and final disposal of the product (Mascarenhas et al. 2016).

EPR imposes manufacturers of electric and electronic equipment with the task of retrieving and recycling their products as soon as it reaches its end of life. The EPR serves as a strategy of saddling manufacturers (rather than the society) of electronic products with the responsibility of bearing the costs associated with managing, recycling and disposing of a particular product (Jaiswal et al. 2015). Since disposal of E-waste, without any obligation on their importer and unauthorized recycler with improper technology, can promote the transboundary flow of these E-wastes, EPR comes in as a strategy to check this menace (Pathak et al. 2017).

EPR was established with backing for the polluter pay principle and the acknowledgement of the importance of improving the management and recycling of waste as agreed at the Rio Earth Summit in 1992 (Nnorom and Osibanjo 2008a, b). The policy instruments introduced under EPR include various types of product fees and taxes, e.g. advance recycling fees (ARFs), product take-back mandates, virgin material taxes, and their combinations. Also included are pay-as-you-throw, waste collection charges, and landfill bans (Pathak et al. 2017).

EPR aims at achieving the following goals:

- Developing electrical and electronic equipment with a 'green' approach and hence constrained the use of components that are hazardous to the environment
- Retraction of products after their end of life as a take-back process
- Recycling and reutilizing of used up products in order to control the generation of E-waste

Although EPR plays an imperative role in defining the tasks and duties associated with E-waste recycling, it does not imply that the burden of E-waste recycling should be left to the manufacturers alone. The alliance and coordination of multi-stakeholders is also a vital element of EPR. Therefore, the government, manufacturers, sellers, mobile telecom carriers, professional recovery operators and consumers should all partake in E-waste recycling (Yin et al. 2014). There is also need for others such as consulting bodies, investment firms and experienced personnel in various localities to collaborate towards ensuring the widespread application of green E-waste treatment technologies by providing capital support and management expertise (Lu et al. 2015).

8.1.2 Mesoscopic Management of E-Waste

The mesoscopic strategy of E-waste is a proactive instead of reactive measure used to effectively manage E-waste. The mesoscopic approaches to E-waste management include material compatibility analysis, life cycle assessment (LCA), material flow analysis (MFA), and multicriteria analysis (MCA).

8.1.2.1 Material Compatibility Analysis

Material compatibility analysis is important to determine if the production chemicals are compatible with the materials of construction of the chemical storage, chemical delivery and production systems (Zeng et al. 2017). This includes both metallic and non-metallic materials.

8.1.2.2 Material Flow Analysis of E-Waste

Several studies have demonstrated the essence of material flow analysis (MFA) as a veritable tool in the management of E-waste. This analysis reveals the genesis to exodus (i.e. reuse, storage, recycling and deposit in landfills) of electronic products (Ikhlayel 2018). MFA is a tool that aims at detailing the route of materials (such as E-wastes) flowing into recycling sites, or disposal areas and stocks of materials, in space and time. MFA links sources, pathways and the intermediate and final termini of the material (Kiddee et al. 2013).

MFA considers the flow of E-waste and its concomitant evaluation in terms of environmental, economic and social values. This analysis is carried out using software-based simulations (Singh 2016). For a successful implementation of MFA, data availability is vital (Kahhat and Williams 2012). Accurate MFA of E-wastes is hindered by a deficiency of accurate data on the quantity of E-waste in a given economy owing to the fact that no record exists for E-waste products in the national statistics of goods produced, sold and traded in (Lau et al. 2013). To a degree, the issue of deficient data can be circumvented by construction using the principle of mass balance (Kahhat and Williams 2012). Hence implementing an inclusive approach in assembling data for MFA could help minimize environmental hazards and maximizing potential resources in a particular system (Agamuthu et al. 2015).

A primary step to be taken for the successful MFA for E-wastes in any country is the creation of an inventory of E-wastes in that region. To accurately do this, sales data (obtained during production, importation, exportation), stock data or quantity of equipment currently in use (determined from households using these devices and/or workplaces where E-waste can be found) and the average lifetime of the items (this depends on the behaviour of each consumer) are required (Lau et al. 2013).

8.1.2.3 Life Cycle Assessment of E-Waste

Life cycle assessment (LCA) can be defined as a systematic strategy that can be used to assess and quantify the environmental performance associated with the various phases of a product creation, processes and activities (Hong et al. 2015). LCA also finds application in defining many environment impact(s) categories such as carcinogens, climate change, the ozone layer, ecotoxicity, acidification, eutrophication and land use, to improve the environmental performance of products (Kiddee et al. 2013).

LCA has also been used to design eco-friendly electronic devices and to curtail E-waste problems (Kiddee et al. 2013). LCA can also be used to systematically evaluate and identify environmental inventory, impact, key factors, decisions, optimization and improvement opportunities associated with all stages of system boundary concurrently (Hong et al. 2015).

LCA can be used to recognize latent environmental impacts to develop eco-design products such as printers, desktop personal computers, heating and air conditioner devices, washing machines and toys.

8.1.2.4 Multicriteria Analysis of E-Waste

Multicriteria analysis (MCA) is a decision-making device developed for considering tactical decisions and resolving multifaceted multicriteria problems that include qualitative/quantitative aspects of the problem (Kiddee et al. 2013). MCA decision-making tool can be used by decision-makers to discover the most appropriate portfolio that consists of multiple measures for overcoming the barriers in the way of sustainable E-waste management under the conditions of uncertainty and incomplete information (An et al. 2015).

8.1.3 *Microscopic Management of E-Waste*

8.1.3.1 Recycling of E-Waste to Recover Valuable Materials

Improved standard of living of citizens, fast economic growth and enhanced technological advancement has resulted in the production of a large amount of electrical and electronic equipment (Lu and Xu 2016). The useful life of electronic and electrical equipment has been reduced over the years owing to the adjustment of consumer taste and technological innovations (Khaliq et al. 2014). This has led to the accumulation of the resultant waste electrical and electronic equipment (WEEE) or E-waste in the environment (Khaliq et al. 2014; Lekka et al. 2015; Akcil et al. 2015; Cayumil et al. 2016; Heydarian et al. 2018). Within this waste stream, the major interests are the printed wiring boards and the plastics (Lu and Xu 2016). Besides environmental concerns, recycling of this waste is attractive and also viable since it contains significant amounts of precious metals (Cayumil et al. 2016; Ebin and Isik 2016). The presence of precious metals such as palladium, platinum, gold, tantalum, selenium, etc., in E-waste, makes recycling a desirable process (Khaliq et al. 2014; Chen et al. 2018). Therefore, their availability is paramount for their inclusion as a tool for economic sustainability (Isıldar et al. 2018). However, it is important to note that E-waste could be hazardous due to the presence of heavy metals (lead, mercury, cadmium, etc.) as well as brominated flame retardants (Kaya 2016). Therefore, proper management options should be adopted to prevent human and environmental health risks (Kaya 2016; Lu and Xu 2016). The technologies that have been successfully deployed in the recovery of metals are hydrometallurgy (use of aqueous solution in extraction of metals from the waste), pyrometallurgy (application of heat to melt appliances and recover metal) and biohydrometallurgy (bioleaching with adapted microorganisms) methods (Cui and Zhang 2008; Chauhan et al. 2018). The following section will briefly discuss these methods.

8.1.3.1.1 Pyrometallurgy for the Recovery of Metal and Energy from E-Waste

Pyrometallurgy involves the recovery of metal of interest by processing (e.g. melting, burning under high temperature) of pulverized E-waste (Cayumil et al. 2016; Ebin and Isik 2016). Most full-scale pyrometallurgical processing of WEEE scrap takes place using smelters designed for refining metals from ore or metal scrap (Townsend 2011; Chauhan et al. 2018). However, the upgrade of final metal products recovered from WEEE is a challenging task due to a mixture of pure metals and alloys after the pyrometallurgical process (Reck and Graedel 2012). Smelting of E-waste could result in the emission of environmentally persistent compounds such as dioxin from the plastic components of the waste. A large amount of slag generation, loss of precious metals and difficulty in recovery of Al, Fe and other metals are the additional problems associated with pyrometallurgical methods (Chauhan et al. 2018). A modified form – vacuum metallurgy – utilizes variation in vapour pressures existing among various metals to improve the selective recovery of desired metals (Townsend 2011). Furthermore, Zhang and Xu (2016) reported an enhanced metal extraction when pyrometallurgical technology was combined with some mild extracting reagent such as ammonia or chloride.

Recently, investigators have focused on energy recovery from E-waste to compensate the high energy demand. This is achieved through pyrolysis of plastic in the presence of suitable catalysts to produce oil with high calorific value when compared to commercial fuel (Sharuddin et al. 2016). Also, plastic components of the E-waste under the influence of catalyst have been shown to form aromatic oil (gasoline) when pyrolyzed (Muhammad et al. 2015). For instance, the plastics from equipment containing cathode ray tubes (CRTs) and also plastic waste from refrigeration equipment have been successfully converted to derivatives of aromatic hydrocarbons. Addition of the Y zeolite and zeolite ZSM-5 to the pyrolysis process resulted in a reduced concentration of styrene, but appreciable concentrations of benzene and its derivatives (toluene and ethylbenzene) were found in the product oil (Muhammad et al. 2015). Using microwave-aided pyrolysis on the plastic fraction of E-waste produced dense and viscous liquid fractions with a high concentration of useful chemicals such as xylenes and styrenes (Rosi et al. 2018).

Though pyrolysis oil could be obtained from the application of pyrolysis technology on WEEE-based plastics, the presence of the brominated flame retardants (BFRs) makes the process problematic (Wang and Xu 2014). Technology that is gasification-based and supercritical fluids methods have been suggested to achieve effective recycling with minimal impact on the environment when compared with a process such as heating at very high temperature. Other components such as glass from the cathode ray tube or liquid display glass can be reutilized sometimes to produce some precious metals such as tin and indium. Nonetheless, before any option can be used for the recycling at a commercial scale, it is very important to conduct an environmental impact assessment to ascertain the effect in the long run (Wang and Xu 2014).

8.1.3.1.2 Hydrometallurgy for Selective Metal Recovery

Compared with pyrometallurgy, hydrometallurgical methods have become desirable treatments for E-waste towards metal extraction/recovery due to reduced gas emission from the latter than the former. Other merits of hydrometallurgy include cost-effectiveness and ease of operation especially under laboratory settings (Chauhan et al. 2015). At a smaller scale, they may present a better control of the processes, thereby ensuring higher efficiency of metal recovery (Chauhan et al. 2018). In this method, alkaline or acidic leaching agents are used to wash the E-waste in order to dissolve and recover the metal of interest. This is accompanied by different forms of physical-chemical methods to finalize metal extraction (Soare et al. 2016). Chauhan et al. (2018) have shown that solvents especially halides, cyanides, thiourea and thiosulfates could be used for the leaching metals from ores.

With a view to averting the unfavourable impacts of smelting process utilized in the recovery of copper from metal powders of waste printed wiring boards rich in tin, Yang et al. (2017) have proposed hydrometallurgy as an effective technique that selectively extracts tin as well as its associated metals. For instance, alkaline pressure oxidation leaching parameters on metal conversion have been systematically investigated. The results showed that Sn, Pb, Al and small amounts of Zn in the metal powders were leached out, leaving a copper residue (Yang et al. 2017). The use of different cyanide or non-cyanide leaching techniques to recover precious and other valuable metals has been reported (Akcil et al. 2015). A novel methodology, using ammonium persulfate ($(NH_4)_2S_2O_8$), to recover gold from waste E-waste has been assayed. The findings presented by Alzate et al. (2016) revealed that $(NH_4)_2S_2O_8$ in aqueous media could be used to recover gold from E-waste. According to Sun et al. (2015), a new electrodeposition process has been proven feasible with high efficiency during copper recovery from E-waste. The leaching solutions have been analysed by ICP in order to detect the most important metals for the electrodeposition (Lekka et al. 2015). The use of hydrometallurgical techniques such as spontaneous reduction polyaniline coating of cotton fibre has been demonstrated as effective tools in gold recovery from electronic scrap (Lekka et al. 2015). Soare et al. (2016) and Popescu et al. (2018) have shown that ionic liquids can be deployed to anionically dissolve E-waste in order to recover metals such as Sn, Pb Au and Ag from multi-component alloy.

8.1.3.1.3 Biohydrometallurgy: An Eco-friendly Approach for Metal Recovery

Biohydrometallurgy has been considered as an important means of metal recovery from waste due to its cost-effectiveness and eco-friendly nature. It is also known as to conserve energy due to its ease in operation compared to other recovery techniques. Through bioleaching and oxidation reactions, different chemolithotrophic bacteria such as *Acidithiobacillus thiooxidans* and *A. ferrooxidans* have been successfully used as bioleaching agents for metal recovery from E-wastes (Chauhan et al. 2018).

An acclimatized consortium of either *Thermoplasma acidophilum* and *Sulfobacillus thermosulfidooxidans* or *S. acidophilus* and *S. thermosulfidooxidans* was used to bioleach more than 75 % of Zn^{2+}, Ni^{2+}, Cu^{2+} and Al^{3+} from E-waste pretreated with iron sulphide and elemental sulphur (Ilyas et al. 2013). Using *A. thiooxidans* (DSM 9463), bioleaching of about 99 % of neodymium, emporium and cerium and 80 % of yttrium and lanthanides from a solution of shredded dust of electronic scraps was accomplished. Also, *Pseudomonas putida* WSC361 (cyanide producer) in a further step sequestered about 45 % Au from shredded dust already bioleached with *A. thiooxidans* (Marra et al. 2018). Priya and Hai (2018) utilized exopolymeric substances produced by *A. ferrooxidans* supplemented with lemon juice to recover nickel, copper, lead and zinc from electronic-waste. Heydarian et al. (2018) demonstrated that *A. ferrooxidans* and *A. thiooxidans* were shown to be effective bioleaching tools for the recovery of cobalt, nickel and lithium from spent batteries of computers (laptops). The mediation of bioleaching of copper from E-waste by metabolites and extracellular enzymes produced by *Acinetobacter* sp. has also been reported (Jagannath et al. 2017). Using *Leptospirillum ferriphilum* as a biolixiviant, Bryan et al. (2015), dissolved metals from printed wiring boards and effectively recovered copper. In spite of various merits of bioleaching, its commercialization has not been attained due to the fact that the process is slow (time-consuming) and selective to particular groups of metals. Also, the microorganisms used in this process are often sensitive to environmental factors such as pH and temperature. Therefore, complete metal recovery using bioleaching technique has not been feasible in the majority of the cases. Thus, there is a need for investigators to develop faster and cheap bioleaching process profitable for pilot-scale operation (Chauhan et al. 2018).

8.1.3.2 Benefits, Challenges and Future of E-Waste Recycling

E-wastes are comprised of many organic heavy metals which although harmful, find great applications in some industries. In designing an efficient system for E-waste recycling, the following factors must be considered: the relevant applicable legislation, the coverage of recycling products, the capital source, the producer responsibility and the effectiveness of the execution of the recycling process (Miao et al. 2017).

There are a number of benefits accruing to the recovery of resources and recycling of E-waste. Recycling E-wastes has also been presented as a lucrative business venture. The recycling could either be formal or informal with the former being the predominant type of recycling in developed nations while the latter being the commonly practised recycling in developing countries (Ramesh Babu et al. 2007).

Major components of most electronic equipment in use today are precious and special metals. This has hence made the manufacture of various electronic products an important contributor to world demand for metals (Sthiannopkao and Wong 2013). Reuse and recycling of metal from E-waste increases metal availability for various products while reducing the dependence on mining industries for the production of new metals, with the subsequent environmental implications of mining activities (Kumar and Holuszko 2016). Recovering these metals from E-wastes paves way for urban mining and hence ensures safe disposal of these hazardous materials for environmental and public safety (Kumar and Holuszko 2016). Moreover, E-waste recovery also enables the extraction of mineral resources much needed in the electronics industry. However, in order to establish this recovery system, it is paramount to 'consummately' analyse the 'environment-resource-cost' balance (Miao et al. 2017).

The major cost constraint in the recycling of E-wastes borders on the collection and transport of the waste. In developing countries, E-wastes are collected by the informal sector (Sthiannopkao and Wong 2013). In order to abate this challenge, manufacturers of electronic products should get involved in the collection and recycling of these waste electronics as a way of incorporating social responsibility (CSR)

(Jaiswal et al. 2015). Available information indicates that there are some advantages when manufacturers of electrical/electronic appliances recycle the waste electronic products. Firstly, manufacturers can easily ascertain the flow of electronic products in the market according to the recycling amount of certain kind of products and quickly grasp the market demand information. Secondly, manufacturers are familiar with the product design process, which gives them an easy workaround when disassembling the waste electronic products and hence saves time and effort and further improves the economic benefit of the recycling process (Miao et al. 2017).

Another obstacle in the proper recovery of precious metals from E-wastes is the fact that E-wastes are frequently illegally moved from developed countries to developing nations lacking infrastructures to recycle such E-wastes and as such recover the target metals. Using low-technological methods to recover metals leads to the loss of valuable metal, thus making the recycling process a futile one (Sthiannopkao and Wong 2013).

In order to enhance the recovery of valuable materials from E-waste, the electronic product has to be disassembled. This is commonly done using crude methods especially in developing worlds. Some developed countries have developed high-tech means of recovering materials from E-wastes (Tsydenova and Bengtsson 2011). However, to improve the ease of disassembling E-wastes, electronic products can be designed so as to allow for ease of disassembly. This can go a long way in making the recovery of a variety of valuable metals from E-wastes less labour-intensive and more cost-effective. Additionally, this will make the process more sustainable by eliminating the need to use chemicals to recover metals which themselves end up as hazardous materials in the environment (Tansel 2017).

Apart from enlightening the public on the need to embrace recycling of E-waste as a green approach to its management, incentives could also be provided in order to further attract the interest of these consumers in bringing recyclable waste electronics forward. Since it has been identified that collection of these waste electronics poses a major challenge in E-waste recycling, getting the consumers to submit used electronics products to either recycling firms or to the manufacturer of such product under a pay scheme will be a boost to overcome this challenge. A transparent and organized platform where consumers bring forward E-wastes for recycling will improve the public awareness on E-wastes and thus help eliminate the rising availability of E-waste in the open environment.

The different methods that have been used to collect E-wastes hitherto include E-waste collection from households or commercial firms through governmental or non-governmental organizations, a collection of electronic scraps from solid waste garbage collectors and collection by the producers of electronic products from end-users (Jaiswal et al. 2015).

In the near future, recycling technology of E-waste would be very vital and attractive sector from the environmental and economic point of view. Recycling converts various E-waste streams to a sustainable and valuable secondary source of metals. In order to be an environmentally sustainable venture, recycling technology should manage E-waste with high efficiency and under reduced carbon footprint (Kaya 2016).

References

Agamuthu P, Kasapo P, Nordin NAM (2015) E-waste flow among selected institutions of higher learning using material flow analysis model. Resour Conserv Recycl 105:177–185. https://doi.org/10.1016/j.resconrec.2015.09.018

Akcil A, Erust C, Gahan CS, Ozgun M, Sahin M, Tuncuk A (2015) Precious metal recovery from waste printed circuit boards using cyanide and non-cyanide lixiviants – A review. Waste Manag 45:258–271. https://doi.org/10.1016/j.wasman.2015.01.017

Alzate A, López ME, Serna C (2016) Recovery of gold from waste electrical and electronic equipment (WEEE) using ammonium persulfate. Waste Manag 57:113–120. https://doi.org/10.1016/j.wasman.2016.01.043

An D, Yang Y, Chaid X, Xi B, Dong L, Ren J (2015) Mitigating pollution of hazardous materials from WEEE of China: Portfolio selection for a sustainable future based on multi-criteria decision making. Resour Conserv Recycl 105:198–210. https://doi.org/10.1016/j.resconrec.2015.10.025

Borthakur A, Govind M (2017) How well are we managing E-waste in India: evidences from the city of Bangalore. Energy Ecol Environ 2(4):225–235. https://doi.org/10.1007/s40974-017-0060-0

Bryan CG, Watkin EL, McCredden TJ, Wong ZR, Harrison STL, Kaksonen AH (2015) The use of pyrite as a source of lixiviant in the bioleaching of electronic waste. Hydrometallurgy 152:33–43. https://doi.org/10.1016/j.hydromet.2014.12.004

Cayumil R, Khanna R, Rajarao R, Mukherjee PS, Sahajwalla V (2016) Concentration of precious metals during their recovery from electronic waste. Waste Manag 57:121–130. https://doi.org/10.1016/j.wasman.2015.12.004

Chauhan G, Pant KK, Nigam KDP (2015) Chelation technology: a promising green approach for resource management and waste minimization. Environ Sci Process Impacts 17:12–40. https://doi.org/10.1039/c4em00559g

Chauhan G, Jadhaob PR, Pant KK, Nigam KDP (2018) Novel technologies and conventional processes for recovery of metals from waste electrical and electronic equipment: Challenges & opportunities – a review. J Environ Chem Eng 6:1288–1304. https://doi.org/10.1016/j.jece.2018.01.032

Chen Z, Niu B, Zhang L, Xu Z (2018) Vacuum pyrolysis characteristics and parameter optimization of recycling organic materials from waste tantalum capacitors. J Hazard Mater 342:192–200. https://doi.org/10.1016/j.jhazmat.2017.08.021

Cui J, Zhang L (2008) Metallurgical recovery of metals from electronic waste: a review. J Hazard Mater 158:228–256. https://doi.org/10.1016/j.jhazmat.2008.02.001

Ebin B, Isik MI (2016) Chapter 5: Pyrometallurgical processes for the recovery of metals from WEEE. In: Alexandre Chagnes, Gérard Cote, Christian Ekberg, Mikael Nilsson, Teodora Retegan WEEE recycling Research, development, and policies Amsterdam: Elsevier 107–137. https://doi.org/10.1016/B978-0-12-803363-0.00005-5

Heydarian A, Mousavi SM, Vakilchapm F, Baniasadi M (2018) Application of a mixed culture of adapted acidophilic bacteria in two-step bioleaching of spent lithium-ion laptop batteries. J Power Sources 378:19–30. https://doi.org/10.1016/j.jpowsour.2017.12.009

Hong J, Shi W, Wang Y, Chen W, Li X (2015) Life cycle assessment of electronic waste treatment. Waste Manag 38:357–365. https://doi.org/10.1016/j.wasman.2014.12.022

Ikhlayel M (2018) An integrated approach to establish e-waste management systems for developing countries. J Clean Prod 170:119–130. https://doi.org/10.1016/j.jclepro.2017.09.137

Ilyas S, Lee J, Chi R (2013) Bioleaching of metals from electronic scrap and its potential for commercial exploitation. Hydrometallurgy 131(132):138–143. https://doi.org/10.1016/j.hydromet.2012.11.010

Isıldar A, Rene ER, van Hullebusch ED, Lens PNL (2018) Electronic waste as a secondary source of critical metals: Management and recovery technologies. Resour Conserv Recycl 135:296–312. https://doi.org/10.1016/j.resconrec.2017.07.031

Jagannath A, Shetty V, Saidutta KMB (2017) Bioleaching of copper from electronic waste using *Acinetobacter* sp. Cr B2 in a pulsed plate column operated in batch and sequential batch mode. J Environ Chem Eng 5(2):1599–1607. https://doi.org/10.1016/j.jece.2017.02.023

Jaiswal A, Samuel C, Patel BS, Kumar M (2015) Go green with WEEE: eco-friendly approach for handling e-waste. Procedia Comput Sci 46:1317–1324. https://doi.org/10.1016/j.procs.2015.01.059

Kahhat R, Williams E (2012) Materials flow analysis of e-waste: domestic flows and exports of used computers from the United States. Resour Conserv Recycl 67:67–74. https://doi.org/10.1016/j.resconrec.2012.07.008

Kaya M (2016) Recovery of metals and nonmetals from electronic waste by physical and chemical recycling processes. Waste Manag 57:64–90. https://doi.org/10.1016/j.wasman.2016.08.004

Khaliq A, Rhamdhani MA, Brooks G, Masood S (2014) Metal extraction processes for electronic waste and existing industrial routes: A review and Australian perspective. Resources 3:152–179. https://doi.org/10.3390/resources3010152

Kiddee P, Naidu R, Wong MH (2013) Electronic waste management approaches: An overview. Waste Manag 33:1237–1250. https://doi.org/10.1016/j.wasman.2013.01.006

Kumar A, Holuszko M (2016) Electronic waste and existing processing routes: A Canadian perspective. Resources **5**(4):35. https://doi.org/10.3390/resources5040035

Lau WK, Chung S, Zhang C (2013) A material flow analysis on current electrical and electronic waste disposal from Hong Kong households. Waste Manag 33:714–721. https://doi.org/10.1016/j.wasman.2012.09.007

Lekka M, Masavetas I, Benedetti AV, Moutsatsou A, Fedrizzi L (2015) Gold recovery from waste electrical and electronic equipment by electrodeposition: A feasibility study. Hydrometallurgy 157:97–106. https://doi.org/10.1016/j.hydromet.2015.07.017

Lu C, Zhang L, Zhong Y, Ren W, Tobias M, Mu Z, Ma Z, Geng Y, Xue B (2015) An overview of e-waste management in China. J Mater Cycles Waste Manag 17:1–12. https://doi.org/10.1007/s10163-014-0256-8

Lu Y, Xu Z (2016) Precious metals recovery from waste printed circuit boards: A review for current status and perspective. Resour Conserv Recycl 113:28–39. https://doi.org/10.1016/j.resconrec.2016.05.007

Marra A, Cesaro A, Rene ER, Belgiorno V, Lens PNL (2018) Bioleaching of metals from WEEE shredding dust. J Environ Manag 210:180–190. https://doi.org/10.1016/j.jenvman.2017.12.066

Mascarenhas O, D'Souza D, George S (2016) Ethics of e-waste management: an input–process–output analytic approach. Manag Labour Stud 41(1):1–18. https://doi.org/10.1177/0258042X16649467

Miao S, Chen D, Wang T (2017) Modelling and research on manufacturer's capacity of remanufacturing and recycling in closed-loop supply chain. J Stat Comput Simul 87(11):1–11. https://doi.org/10.1080/00949655.2017.1366488

Muhammad C, Onwudili JA, Williams PT (2015) Catalytic pyrolysis of waste plastic from electrical and electronic equipment. J Anal Appl Pyrolysis 113:332–339. https://doi.org/10.1016/j.jaap.2015.02.016

Nnorom IC, Osibanjo O (2008a) Electronic waste (e-waste): Material flows and management practices in Nigeria. Waste Manag 28:1472–1479. https://doi.org/10.1016/j.wasman.2007.06.012

Nnorom IC, Osibanjo O (2008b) Overview of electronic waste (e-waste) management practices and legislations, and their poor applications in the developing countries. Resour Conserv Recycl 52:843–858. https://doi.org/10.1016/j.resconrec.2008.01.004

Pathak P, Srivastava R, Ojasvi (2017) Assessment of legislation and practices for the sustainable management of waste electrical and electronic equipment in India. Renew Sust Energ Rev 78:220–232. https://doi.org/10.1016/j.rser.2017.04.062

Popescu AM, Yanushkevich K, Soare V, Donath C, Neacsu EI, Constantin V (2018) Recovery of metals from anodic dissolution slime of waste from electric and electronic equipment (WEEE) by extraction in ionic liquids. Chem Res Chin Univ 34(1):113–118. https://doi.org/10.1007/s40242-017-7225-4

Premalatha M, Tasneem A, Abbasi T, Abbasi SA (2014) The generation, impact, and management of e-waste: state of the art. Crit Rev Env Sci Tec 44 (14):1577–1678. https://doi.org/10/1080/10643389.2013.782171

Priya A, Hai S (2018) Extraction of metals from high grade waste printed circuit board by conventional and hybrid bioleaching using *Acidithiobacillus ferrooxidans*. Hydrometallurgy 177:132–139. https://doi.org/10.1016/j.hydromet.2018.03.005

Ramesh Babu B, Parande AK, Basha CA (2007) Electrical and electronic waste: A global environmental problem. Waste Manag Res 25:307–318. https://doi.org/10.1177/0734242X07076941

Reck BK, Graedel TE (2012) Challenges in metal recycling. Science 337:690–695. https://doi.org/10.1126/science.1217501

Rosi L, Bartoli M, Frediani M (2018) Microwave assisted pyrolysis of halogenated plastics recovered from waste computers. Waste Manag 73:511–522. https://doi.org/10.1016/j.wasman.2017.04.037

Sharuddin SDA, Abnisa F, Daud AWM, Aroua MK (2016) A review on pyrolysis of plastic wastes. Energy Convers Manag 115:308–326. https://doi.org/10.1016/j.enconman.2016.02.037

Singh K (2016) E-waste management and public health: A scenario of Indian cities. Int J Sci Res Sci Enginer Technol 2(3):887–890. https://doi.org/10.4103/0019-5278.38461

Soare V, Dumitrescu D, Burada M, Constantin I, Soare V, Capota P, Popescu AM, Constantin V (2016) Recovery of metals from waste electrical and electronic equipment (WEEE) by anodic dissolution. Rev Chim 67(5):920–924. www.revistadechimie.ro

Sthiannopkao S, Wong MH (2013) Handling e-waste in developed and developing countries: Initiatives, practices, and consequences. Sci Total Environ 463(464):1147–1153. https://doi.org/10.1016/j.scitotenv.2012.06.088

Sun Z, Xiao Y, Sietsma J, Agterhuis H, Yang Y (2015) A cleaner process for selective recovery of valuable metals from electronic waste of complex mixtures of end-of-life electronic products. Environ Sci Technol 49(13):7981–7988. https://doi.org/10.1021/acs.est.5b01023

Tansel B (2017) From electronic consumer products to e-wastes: Global outlook, waste quantities, recycling challenges. Environ Int 98:35–45. https://doi.org/10.1016/j.envint.2016.10.002

Townsend TG (2011) Environmental issues and management strategies for waste electronic and electrical equipment. J Air Waste Manag Assoc 61:587–610. PMID: 21751577

Tsydenova O, Bengtsson M (2011) Chemical hazards associated with treatment of waste electrical and electronic equipment. Waste Manag 31:45–58. https://doi.org/10.1016/j.wasman.2010.08.014

Wang R, Xu Z (2014) Recycling of non-metallic fractions from waste electrical and electronic equipment (WEEE): A review. Waste Manag 34(8):1455–1469. https://doi.org/10.1016/j.wasman.2014.03.004

Yang T, Zhu P, Liu W, Chen L, Zhang D (2017) Recovery of tin from metal powders of waste printed circuit boards. Waste Manag 68:449–457. https://doi.org/10.1016/j.wasman.2017.06.019

Yin J, Gao Y, Xu H (2014) Survey and analysis of consumers' behaviour of waste mobile phone recycling in China. J Clean Prod 65:517–525. https://doi.org/10.1016/j.jclepro.2013.10.006

Zeng X, Song Q, Li J, Yuan W, Duan H, Liu L (2015) Solving e-waste problem using an integrated mobile recycling plant. J Clean Prod 90:55–59. https://doi.org/10.1016/j.jclepro.2014.10.026

Zeng X, Duan H, Wang F, Li J (2017) Examining environmental management of e-waste: China's experience and lessons. Renew Sust Energ Rev 72:1076–1082. https://doi.org/10.1016/j.rser.2016.10.015

Zhang L, Xu Z (2016) A review of current progress of recycling technologies for metals from waste electrical and electronic equipment. J Clean Prod 127:19–36. https://doi.org/10.1016/j.jclepro.2016.04.004

Zhang Q, Ye J, Chen J, Xu H, Wang C, Zhao M (2014) Risk assessment of polychlorinated biphenyls and heavy metals in soils of an abandoned e-waste site in China. Environ Pollut 185:258–265. https://doi.org/10.1016/j.envpol.2013.11.003

Zhang S, Ding Y, Liu B, Pan D, Chang C, Volinsky AA (2015) Challenges in legislation, recycling system and technical system of waste electrical and electronic equipment in China. Waste Manag 45:361–373. https://doi.org/10.1016/j.wasman.2015.05.015

Chapter 9
Recycling Processes for the Recovery of Metal from E-waste of the LED Industry

Emanuele Caroline Araújo dos Santos, Tamires Augustin da Silveira, Angéli Viviani Colling, Carlos Alberto Mendes Moraes, and Feliciane Andrade Brehm

Contents

Abstract Increasingly used today, the light-emitting diode (LED) technology today replaces other technologies and has gained a notable market share. This growth in use implies an increased demand for specific materials used in LED manufacturer, aiming at improved performance of devices. However, most materials used in LED manufacture are considered critical in terms of availability, since they are increasingly sought after by the industry. Chemical elements like gallium (Ga) and indium (In), rare earth elements like yttrium (Y) and cerium (Ce), and precious metals such as gold (Au) and silver (Ag) are used in LED devices. An additional difficulty concerns the methods used to sort and reuse these materials, especially due to the small amounts used. This poses a considerable challenge in the full recycling of LED devices. Research is carried out to develop sorting and recovery methods for critical metals generated during the production of LED devices and at the

E. C. A. dos Santos · T. A. da Silveira · A. V. Colling
C. A. M. Moraes (✉) · F. A. Brehm
Universidade do Vale do Rio dos Sinos – UNISINOS, Porto Alegre, RS, Brazil
e-mail: cmoraes@unisinos.br

A. Khan et al. (eds.), *E-waste Recycling and Management*, Environmental Chemistry for a Sustainable World 33,
https://doi.org/10.1007/978-3-030-14184-4_9

end of life of these devices. Some of the most important methods developed for this purpose include pyrometallurgical (pyrolysis), hydrometallurgical (acid leaching), and biotechnological technologies (microbial leaching).

Keywords LED industry · LED elements/metals · Recovery chain · Recycling · Recovery/reclamation process · E-waste

9.1 Introduction

Originally used as an indicator light, the light-emitting diode (LED) is an increasingly used technology today, ubiquitously present in homes, industries, and products. LEDs consume comparatively less energy than other devices, at the same time that they have a longer useful life. LEDs are found in several electrical and electronic devices, lamps, bulbs, and even automobiles.

LEDs have been known for over half a century and were originally used in simple applications such as indicator lights in general (Teixeira et al. 2016). However, a new era in the lighting industry dawned with the development of blue LEDs, which have longer life cycles and are produced at reduced costs. Today, blue LEDs have replaced conventional lighting devices (Teixeira et al. 2016).

A multinational bulb company estimates that lighting holds a 20% share of the global energy market, or 2651 TWh/year, of which 70% is consumed by poor efficiency bulbs. As solid-state devices, LEDs are developed based on the properties of inorganic semiconductors that emit light by electroluminescence. Apart from the fact that LEDs do not include mercury, these devices are five times as efficient compared with incandescent lights. Another advantage is that they have longer useful lives (OSRAM 2009).

Electrical and electronic equipment (EEE) is defined as a device that requires electrical current or a magnetic field to work or that generates, transfers, or converts electrical current into a magnetic field. In turn, waste electrical and electronic equipment (WEEE) includes all products, parts, and components of electrical and electronic equipment post-use (Carvalho and Xavier 2014). Therefore, LED bulbs and lighting fixtures fit in the electrical and electronic equipment category since they need a printed circuit board (PCB). These lighting devices are considered lighting electronic equipment in Annex I, Directive 2012/19/EU.

LED waste is formed by several materials, several of which contain a variety of toxic substances that may contaminate the environment and pose hazards to human health if not properly managed (Kiddee et al. 2013). However, if on the one hand waste electrical and electronic equipment represents a considerable environmental liability, on the other they stand as a commercial alternative whose reverse logistics has to be investigated (Carvalho and Xavier 2014).

The evolution of LED technology has led to a longer useful life of these devices, reducing the demand due to replacement. However, the quality of the original product and the use customers make of it affect the actual useful life of LEDs (Gassmann et al. 2016). Quality of these products may be affected by price adjustment due to the fact that the lighting market is fragmented, characterized by marked completion of several small companies. This influences the quality of heat transfer for cooling and on the quality of electric components of the driver (Gassmann et al. 2016).

In addition, it should be considered that small amounts of critical elements like gallium, indium, yttrium, cerium, gold, and silver are used in the manufacture of LEDs. The low amounts of these elements used and the wide variety of other materials used to manufacture one component or product pose a considerable challenge in the recycling and recovery of these materials. Research and technologies that vary from pyrolysis to bacterial leaching have been used in the effort to recover gallium and indium from waste generated in the LED production and in at the end of the useful life of these devices.

This chapter presents a history of LEDs and of the manufacture of these devices, discusses the main materials used, and reviews the advancements in the recovery of these materials in industrial waste and post consumption.

9.2 History of LED Production

The discovery of semiconductors paved the way for the development of electronic devices such as diodes and transistors. Briefly, a diode has two electrodes for terminals, and, although it is the simplest device based on semiconductors, it plays an essential role in electronic systems. In turn, diodes enabled the development of transistors and integrated circuits (Gois 2008; Callister and Rethwisch 2013).

Also called rectifier, a diode is an electronic device that releases the current in one direction only. This means that a diode may be used to convert an alternating current in direct current, for example. A diode works based on a “p-n junction”, which is the junction of two different materials. It is built using a semiconductor part that is doped: d-doping is applied on one side, and p-doping is applied on the other side of the part (Callister and Rethwisch 2013; Castro 2013).

Light-emitting diodes (LEDs) are semiconductor devices built using a p-n junction that emits light when charged with electric current (Fig. 9.1). In other words, a LED is a very small electroluminescent semiconductor chip made of a solid crystalline material measuring 0.25 mm^2 on average (Gois 2008; Ascurra 2013; Callister and Rethwisch 2013). Since this technology does not rely on filaments or gases, it may also be called solid-state lighting (SSL) (Ascurra 2013).

After several attempts to build a light source based on electrical energy, Humphry Davy developed the first bulb using a platinum filament in 1802. But it was only in 1879 that Thomas Edison built the first commercially feasible electrical incandescent light bulb using a very thin coal filament. Today, incandescent light bulbs are produced using a tungsten filament (Castro 2013).

In 1926 Edmund Germer obtained uniform white light from a mercury vapour bulb. It was built using a pressurized glass bulb covered with a phosphorescent powder. In the mid-1930s, the first fluorescent bulbs were commercialized (Castro 2013).

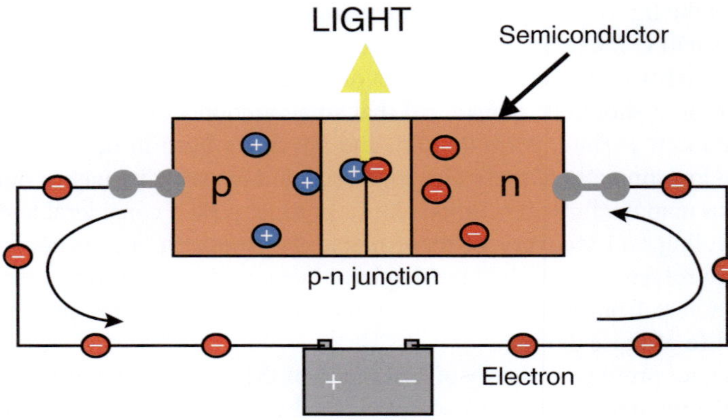

Fig. 9.1 Light emission from a LED. (Source: Gois 2008)

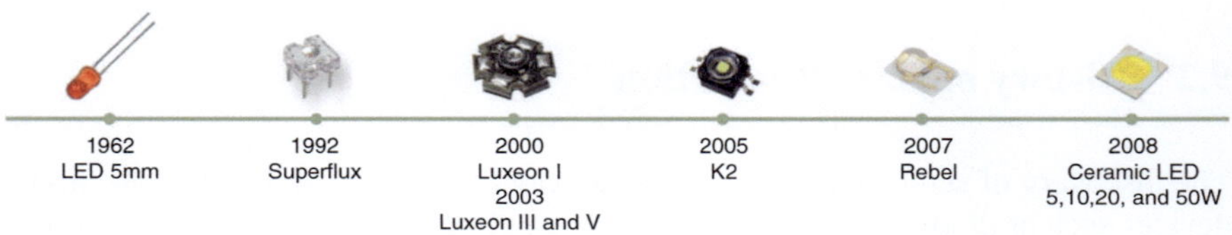

Fig. 9.2 The evolution of LEDs. (Source: Gois 2008)

In 1907, the researcher Henry Joseph Round described an interesting phenomenon, in which a small voltage run across a silicon carbide (SiC) crystal-induced the emission of yellow light. The first LED that emitted visible light (red) using the technology called gallium arsenide phosphide (GaAsP) was developed by a researcher at General Electric named Nick Holonnayak Jr. in 1962. In 1993, the researcher Dr. Shuji Nakamura introduced the first blue high brightness LED, enabling the development of the white LED. The first LED bulbs produced on large scale were introduced in exhibitions in the United States of America and Europe in 1997 and 1998 (Gois 2008). The technology evolved rapidly, at impressive growth rates (Castro 2013). Figure 9.2 shows the evolution of LEDs to the year 2008.

By comparison with conventional lighting (incandescent and fluorescent bulbs), the advantages of LEDs include longer lifespan and higher energy efficiency. LEDs consume 75% less energy than incandescent and fluorescent bulbs (Jang et al. 2014).

Even though it was originally used only as indicator lights in electrical and electronic equipment, LEDs are currently employed in an increasing number of applications, with a growing market share and considerable effects on several economy sectors (Schubert 2006; Dias 2012). The drop in prices and the surprising evolution of this technology afford to launch high-efficiency products and improve adoption by customers. The efficiency and light quality of LEDs have been substantially improved since they were first introduced. In addition, LEDs are mercury-free and may be

Table 9.1 Led types, characteristics, and examples

Type	Characteristics	Image
Indicator LEDs	The oldest kind, used in electronic equipment to indicate on/off; include a colour cap that works as an optical filter to define colour	
High brightness LEDs	Known as HB-LEDs, they present higher luminous flux and are more efficient than indicator LEDs; include a transparent cap, since they are manufactured in colours; used in traffic lights, bus signage, and other applications	
High power LEDs	Operate at higher currents, compared with the types above; operate at power values ≤1 W, requiring heat exchanger, producing high luminous flux; used in indoor lighting and architectural fittings, among other applications	

Source: Adapted from Dias (2012)

designed in a variety of ways, two important advantages that are not afforded by other lighting technologies (Ascurra 2013; Jang et al. 2014; Gassmann et al. 2016).

Since 2007, LED bulbs are available in the market, and currently it is possible to choose between several models of different shapes and socket types. The market offers not only LED bulbs that may be used to replace conventional ones but also several fixture types with their own inbuilt LED module. This wide diversity of products makes it difficult to quantify and qualify the materials characteristically used in LED lighting devices (Gassmann et al. 2016).

There are three types of LEDs. Table 9.1 shows LED types and respective characteristics.

According to Dias (2012), the first LED devices had very low luminous flux and were used only in electronic equipment and signage. It was only after considerable research that LEDs that could operate at 1 W or more were developed, affording a wider array of applications. It was due to high power LEDs that high luminous flux values could be obtained with wider viewing angles that spread light more efficiently, which enabled the use of these LEDs as lighting alternatives. These LEDs replace conventional bulbs in several applications (Dias 2012).

The Brazilian Lighting Industry Association (ABILUX) defined LED as an electrical and electronic device that generates visible light and contains, in addition to LEDs, a driver (an energy source). It replaces conventional bulbs, such as incandescent and fluorescent ones, and therefore is usually designed in similar shapes as those. In addition, LEDs may be manufactured including a screw base or a base that may fit simpler socket types (ABILUX 2017). Figure 9.3 shows examples of LED bulbs. LED lighting fixtures are equipped with several LEDs connected in series or in parallel as a means to increase luminous flux. These LEDs usually present the same characteristics (Dias 2012).

Since LEDs require low voltage, they cannot be connected directly to electrical lines, requiring instead a voltage step-down system or a source (driver) to adjust

Fig. 9.3 Types of LED bulbs. (Source: ABILUX 2017)

voltage values and provide constant current (Dias 2012; ABILUX 2017). However, the fact that LEDs may operate at low voltages and currents means that they do not require a reactor (as opposed to fluorescent lights); therefore, the circuits used to switch LEDs on are simpler (Oliveira 2007; Dias 2012).

In the effort to develop products with better lighting output, high power values are used in LEDs. Nevertheless, the heat generated in the process increases substantially, degrading phosphor, reducing brightness, affecting colour, and decreasing life cycle and luminous efficiency of these products (Castro 2013; Jang et al. 2014). Compact LED fittings such as retrofit bulbs normally require a cooling device as a means to transfer heat to the environment. This prevents overheating, maintaining operation temperatures at ideal values. This cooling device is the largest element in a LED fitting, and it usually is produced using aluminium or a ceramic material that dissipates heat (Castro 2013; Gassmann et al. 2016).

Table 9.2 Parts of a LED bulb

Major parts	Subparts
Base	Insulator
	Contact board
	Plastic cap
	Aluminium board
	Electronic reactor
Bulb	Bulb material
	Heat exchanger
Filling	LEDs

Source: Adapted from Osram (2009)

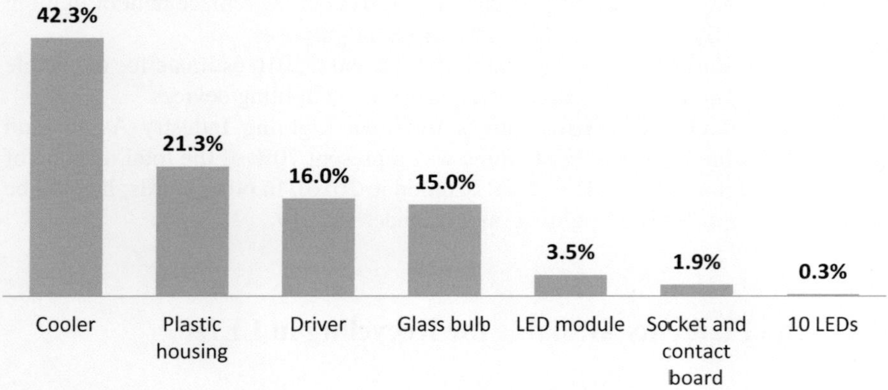

Fig. 9.4 Percent weight composition of a 9.5 W LED bulb with an E27 base. (Source: Adapted from Gassmann et al. 2016)

Osram (2009) divided LED bulbs into three main parts: base, bulb, and filling. These parts were also divided into subparts (Table 9.2).

Gassmann et al. (2016) obtained the per cent weight values of the main parts of a last-generation LED bulb (Fig. 9.4). The product analysed weighed 85.5 g. It was composed of an aluminium cooler, a glass bulb, a plastic housing, a PCB, an E27 socket with a contact board, and a LED module with 10 LEDs fixed on an aluminium board. Together, the ten LEDs weighed 275 mg, accounting for 0.32% of the total weight (Gassmann et al. 2016).

As mentioned above, LED bulbs are manufactured in shapes that resemble conventional bulbs; that is, tubular LED bulbs are used to replace fluorescent tubes, while bulb-shaped LEDs are used to replace compact incandescent and fluorescent ones. For this reason, they are built with the same mechanism of contact with wiring. For example, a bulb-shaped LED includes an Edison screw, similarly to its predecessors, the incandescent and compact fluorescent bulbs. For Yamachita et al. (2006), this part plays the role of attaching the bulb mechanically, providing the electrical connection with the wiring. These parts are usually manufactured in aluminium alloys and include a brass pin (Polanco 2007).

9.3 The LED Industry and the Prospects of Consumption

For Bastos (2011), several regions in the world have banned incandescent bulbs in the effort to improve energy efficiency, following recommendations by the International Energy Agency. According to Teixeira et al. (2016), the LED market share may be divided into four segments based on applications: residential, commercial, industrial, and outdoor (e.g. which includes public lighting and lighting in large open spaces like stadiums and parking lots). Table 9.3 shows the main drivers and critical points discussed by Teixeira et al. (2016) in the replacement of lighting technologies in the adoption of LEDs for different purposes.

Figure 9.5 illustrates the value chain in 2010 and a 2010 estimate for the whole production and commercialization processes of LED lighting devices.

The Fecomércio SP website quotes Brazilian Lighting Industry Association (ABILUX) to predict that LED fixtures will represent 70% of the total revenue of the lighting sector worldwide in 2020 (Comércio 2016). In other words, it could be said that LEDs may be the lighting market leader in 2019.

9.4 The Elements Available for Recycling in LEDs

The combination of semiconductors defines the colour of LEDs. Since the mechanism of LEDs is based on energy levels, a voltage applied to a LED device excites LED crystals to a higher energy level. When these electrons return to their baseline levels, they emit waves of characteristic wavelengths as light. Since each element has its own energy level, which therefore generates different wavelengths, the material used determines the colour of LED (Denbaars 1997 and Cervi 2005; Bullough 2003; Gois 2008).

Table 9.3 Drivers and critical points in the adoption of LEDs for various applications

Application	Main lighting technology	Driver for replacement	Critical points of the LED technology
Residential	Incandescent decoration	Price	Price of bulbs
Commercial	Fluorescent incandescent	Price, durability	Vary with incentive policies and green certification of buildings
Industrial	HID (sodium/mercury vapour) fluorescent	Maintenance costs	Durability
Outdoor	HID (sodium/mercury vapour)	Maintenance costs	Durability

Source: Adapted from Teixeira et al. (2016)

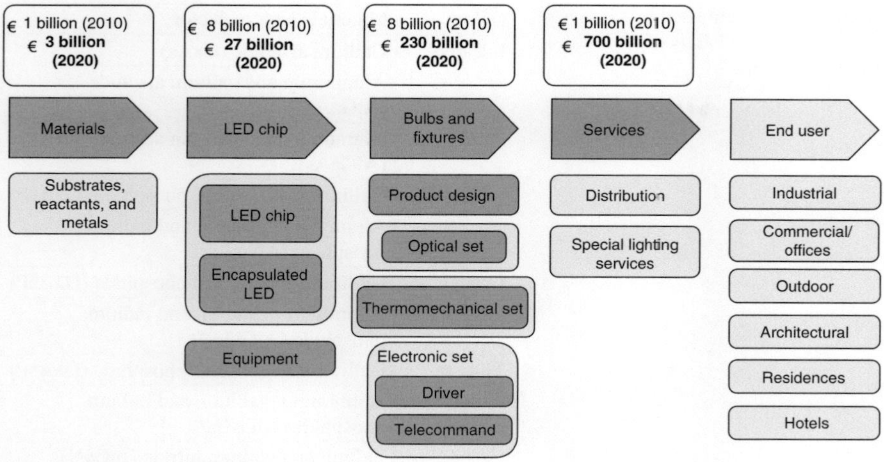

Fig. 9.5 Value chain of LED lighting. (Source: Adapted from Teixeira et al. 2016)

According to Denbaars (1997) and Cervi (2005), the elements usually employed to dope LEDs are gallium, aluminium, arsenic, phosphor, indium, and nitrogen, in addition to a variety of combinations thereof, which determine the colour emitted by these devices. Therefore, each LED is defined by the elements that compose it (Gois 2008). Table 9.4 illustrates the most common elements used in the production of LEDs.

In lighting systems, the main LEDs used are combinations of gallium and indium nitrides (InGaN) to generate blue and green tones and indium, gallium, and aluminium phosphides to generate red, orange, and yellow light (Bullough 2003; Cervi 2005; Schubert 2006). The difference in colours obtained is due to small variations in the proportions of these elements (Bullough 2003; Cervi 2005).

White light LEDs are built according to three main methods. The first, which is also the simplest and most popular, is based on a layer of phosphor applied on a blue LED chip (Goiss 2008; Castro 2013; Gassmann et al. 2016). The second is the RGB method (red, green, blue), which consists of combining red, blue, and green LEDs (Gois 2008; Castro 2013). The third is based on ultraviolet LED used in combination with the RGB (red, green, blue) LEDs (Gois 2008; Dias 2012).

White LEDs, which are used in lighting bulbs, are built using a blue semiconductor chip and a phosphor layer. Gassmann et al. (2016) underscore the fact that a LED may contain small amounts in the microgram range of rare earth elements such as europium and cerium. For example, a 1 mm^2 LED may include 3 μg of cerium or europium. Other rare elements that are found at between 90 μg and 200 μg in 1 mm^2 include yttrium, lutetium, and gadolinium in grenades such as aluminium and yttrium (YAG), aluminium and lutetium (LuAG), and aluminium and gadolinium (GdAG) grenades. In addition, technological metals like gallium and indium are used in the production of blue LEDs (between 17 μg and 25 μg of gallium and 28 ng of indium). Silver, tin, nickel, titanium, silicon, and germanium are also used; gold

Table 9.4 Element used in the production of LEDs of different colours

Colour	Material semiconductor
Infrared	Gallium arsenide (GaAs)
	Aluminium and gallium arsenide (AlGaAs)
Red	Aluminium and gallium arsenide (AlGaAs)
	Gallium arsenide and phosphide (GaAsP)
	Aluminium, gallium, and indium phosphide (AlGaInP)
Orange	Gallium arsenide and phosphide (GaAsP)
	Aluminium, gallium, and indium phosphide (AlGaInP)
Yellow	Gallium arsenide and phosphide (GaAsP)
	Aluminium, gallium, and indium phosphide (AlGaInP)
Green	Indium and gallium nitride (InGaN)/gallium nitride (GaN)
	Gallium phosphide (GaP)
	Aluminium, gallium, and indium phosphide (AlGaInP)
	Aluminium and gallium phosphide (AlGaP)
Blue	Zinc selenide (ZnSe)
	Indium and gallium nitride (InGaN)
	Silicon carbide (SiC) as substrate
Violet	Indium and gallium nitride (InGaN)
Ultraviolet	Diamond (C)
	Aluminium nitride (AlN)
	Aluminium and gallium nitride (AlGaN)
	Aluminium, gallium, and indium nitride (AlGaInN)
White	Blue chip or phosphor ultraviolet (UV)

Source: Adapted from Gois (2008)

is employed in connection wires (approximately 200 mg per diode) (Gassmann et al. 2016). Figure 9.6 presents a microphotograph of a LED showing the overlapping of elements detected by X-ray fluorescence.

Figure 9.7 shows the estimated concentration of rare earth elements in 50 LED bulbs.

The materials and elements used in the production of LEDs are considered critical, especially rare earth elements like lutetium, cerium, and europium, or technological metals like gallium and indium and precious metals such as gold and silver (Gassmann et al. 2016).

According to Ayres and Pieró (2017), critical materials are geologically scarce, affecting commercial availability. In addition, these materials are used for very specific

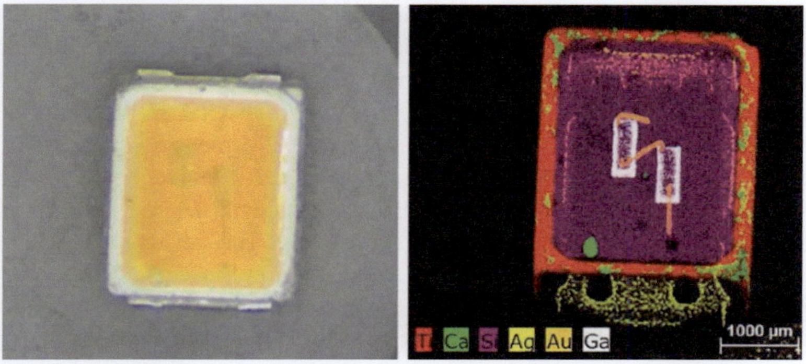

Fig. 9.6 Microphotograph of a LED. (Source: Gassmann et al. 2016)

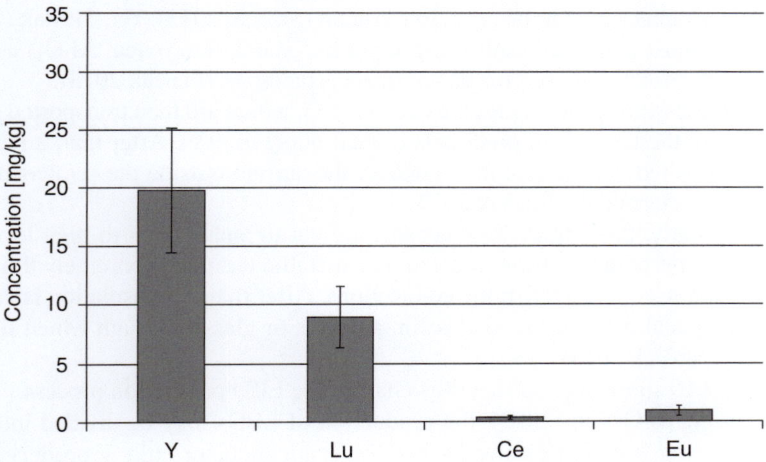

Fig. 9.7 The estimated concentration of rare earth elements in 50 retrofit bulbs. (Source: CycLED 2017)

purposes and are not easily replaced due to their particular properties. More critical rare elements such as gallium, germanium, and indium, besides some rare earth elements, have unique properties, which makes them especially important in the electronics industry. Moreover, all phosphors of different luminous flux values used in the lighting industry include rare earth elements (Ayres and Pieró 2017).

9.5 The LED Production Chain

Kitai (2011) described the most common method used to produce LEDs, the metal-organic vapour phase epitaxy (MOVPE), which may also be called metal-organic chemical vapour deposition (MOCVD). In this method, molecules formed by atoms of semiconductor elements are made to run over a previously heated substrate using an inert gas flow, usually argon. These molecules are formed by organic elements, like hydrogen and carbon, which are bound to one of the atoms required for doping (aluminium, gallium, indium, phosphor, and others) (Kitai 2011). This labour-intensive process is based on the use of a carrier gas formed by a mixture of inert molecules. The substrate should be hot enough to break molecules and afford the deposition of the atoms of the semiconductor elements desired. The remaining atoms of the mixture flow with the gas (Kitai 2011). The advantages of this technique include the possibility to control composition and to work with ternary and quaternary semiconductors, which affords to obtain a wide variety of compositions. Besides, the method allows changing composition during deposition by merely altering the flow rate of molecules (Kitai 2011).

The LED production process starts with the manufacture of ingots (Fig. 9.8), which are used to produce the substrate. The composition of the substrate may vary, since various materials can be used, like silicon (Stasiak 2013). Yet, Stasiak (2013) claims that most LEDs are built using sapphire (aluminium oxide, Al_2O_3) as substrate. Another material used for this purpose is diamond (Stasiak 2013).

From these ingots, fine blades are cut (wafers), which are then transported to the hoods where the deposition process is carried out (Fig. 9.9). After that, a metallic layer is deposited, whose role is to conduct the current outside the semiconductor using metallic terminals (Stasiak 2013).

Subsequently, the wafers are mechanically cut to small chips to high levels of accuracy. These chips are then soldered on a metallic terminal. Extremely thin gold filaments are used to establish the connections. After that, encapsulation is carried out using optical materials (epoxy resin, silicone, or glass), through which light is irradiated (Stasiak 2013).

Figure 9.10 shows a simplified flowchart of the LED production process.

According to Osram (2009), the production of LEDs may be divided into two stages; the first is called frontend when the semiconductor chip is produced; the second is called backend when contacts are connected and the chip is encapsulated. In turn, Scholand and Dillon (2012) divided the process into three main steps: production of the substrate, production of the LED chip, and assembly and encapsulation.

Fig. 9.8 Sapphire ingots. (Source: Stasiak 2013)

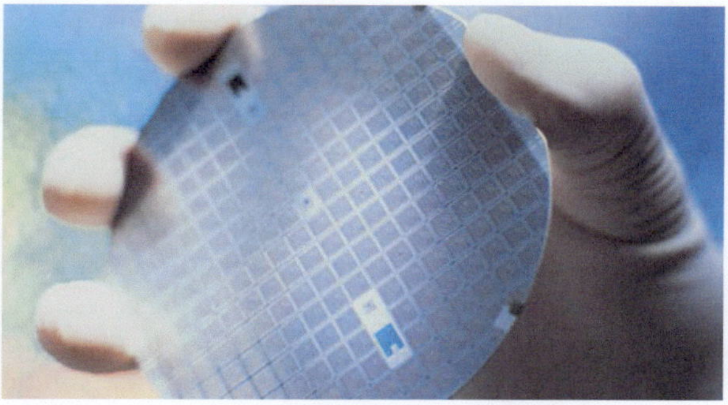

Fig. 9.9 Wafer after deposition of layers. (Source: ABILUX 2017)

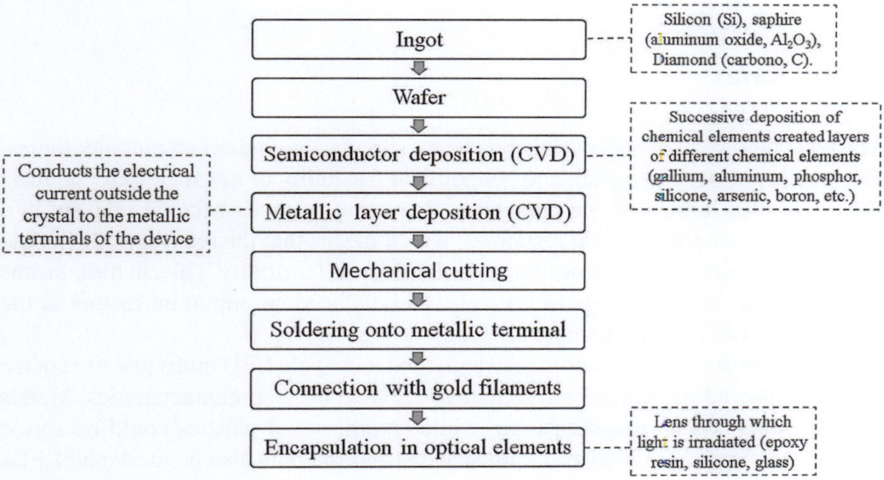

Fig. 9.10 Flowchart of LED production. (Source: Adapted from Stasiak 2013)

9.6 LED Recycling Processes

According to Ayres and Pieró (2017) and Buchert et al. (2009), the amounts of critical metals in finished products are very low, which makes it difficult to classify and sort these metals at the end of a product's lifecycle.

As a circular economy method, the mass recycling of waste is relatively simple, but it poses challenges when recycling modern, complex devices built with several materials (Reuter and Van Schaik 2015).

LEDs present low return rates due to their long life cycle; so, the fate of LED waste is not an urgent issue. However, the fast expansion of the market of LEDs used for lighting makes it clear that the recycling of these products will become environmentally relevant in the future (Gassmann et al. 2016).

Studies on the recovery of materials have discussed several techniques, such as:

- Acid leaching to recover powder gallium and indium generated during the metal-organic chemical vapour deposition (MOCVD) process
- Pyrolysis, physical degradation, and vacuum separation to recover gallium and indium generated from LED chips
- Bacterial leaching to recover gallium from gallium nitride (GaN)
- Bacterial leaching to recover gallium from gallium arsenide (GaAs)
- Mechanical oxidation and leaching of the powder produced during metal-organic chemical vapour deposition (MOCVD) to recover gallium
- Leaching, milling, and annealing to recover gallium from gallium nitride (GaN) from powder waste generated during the manufacture of LEDs

9.6.1 Bulbs

The recyclability of materials depends on existing technologies and methodologies. But the economic, geological, and geopolitical feasibility of raw materials and market prices also influences the parameter. Recycling methods become relevant in a scenario of dwindling natural resources, which means that the recovery of even the smallest fractions will be a question of technological feasibility. This, in turn, means that the amounts and storage of critical materials become important factors in the recycling of LEDs (Gassmann et al. 2016).

Gassmann et al. (2016) introduced a method to recycle LED bulbs that use coarse milling followed by segregation of materials based on their characteristics. Metals may be separated by magnetization, while ceramics and plastics could be sorted using the difference in density. Grain size distribution may also be used, and LEDs are easily detected by irradiation with ultraviolet (UV) light. The electrical and electronic components may be transported to specialized recycling organizations that recover precious metals. The authors also concluded that LEDs could be considered impurities that may be collected and stored, while no suitable method is available to recover critical materials. Due to the small size, LEDs do not require much storage space (Gassmann et al. 2016). The flow proposed by the authors is shown in Fig. 9.11.

9.6.2 Other LED Devices

LEDs are produced as a wide variety of items, such as superthin TV sets, cell phones, lighting fixtures, outdoors, and laptops.

TV sets and cell phones are complex devices when it comes to recycling since they present a large number of different materials associated with LEDs. Therefore, these devices are considered complex waste, for which no technique has been developed to recycle their components. In addition, these products generate considerable amounts of waste, mainly due to the fast evolution of technology.

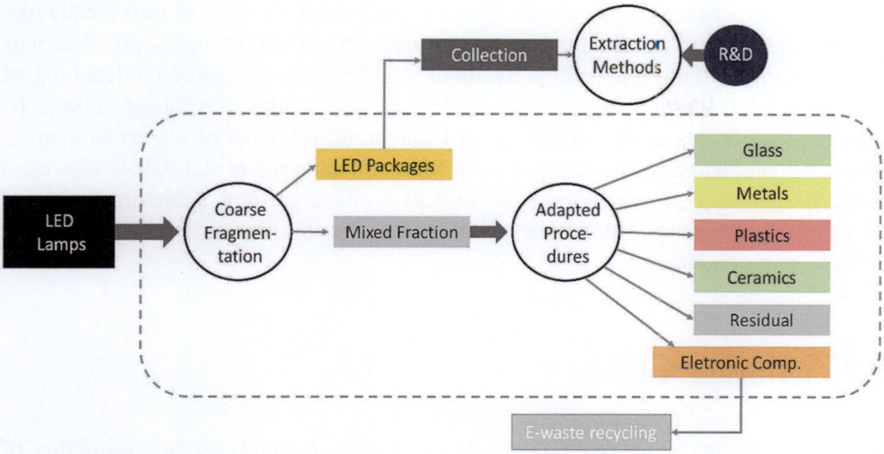

Fig. 9.11 Recycling flow for LED lights. (Source: Adapted from Gassmann et al. 2016)

Currently, there are no established techniques to recycle these devices, and mechanical milling is the most common method employed. However, the best part of valuable elements is concentrated in specific components of these devices. For this reason, milling of these devices does not afford good recovery rates of these elements.

For the Fraunhofer Institute, the removal of the most valuable components enables a differentiated treatment, which in turn allows obtaining higher amounts of recycled material. However, one of the difficulties in this process is a long time required to extract components manually. For example, it has been estimated that 3–4 min is necessary to extract LEDs from notebooks. For TVs, this time was measured to be between 6 and 7 min. Therefore, mechanical disassembly processes are very important concerning efficient recycling management.

The Fraunhofer Institute developed a program financed by the European Commission to recycle elements in LEDs. The method consisted of the mechanical separation of components without destroying LEDs (CEMPRE 2018). Initially, the material is milled and segregated by electrical and hydraulic separation, which enables to break the encapsulation without damaging the LED. Therefore, the components may be recycled individually, to maximum recovery. This technique may also be used in the various LED devices. One of the aims of the program is to improve the recycling of metals in production waste, increase the lifespan of products, minimize waste during manufacture, and promote ecological innovation. In other words, the program aims not only to recycle devices at the end of their life cycle but also to reduce production waste, using minimum amounts of elements to the highest efficiency.

Although it is possible to recycle 98% of a LED device, many countries have not taken any measures to recycle these products (CEMPRE 2018). However, studies have been carried out to investigate the recycling of important elements, especially gallium and indium. Zhan et al. (2015) recovered about 93–95% of rare metals like gallium and indium in LEDs using a new process based on pyrolysis, physical segregation, and vacuum metallurgy. Swain et al. (2015a) investigated the leaching of gallium in LED waste using heterotrophic bacteria. The main advantage of biotechnological processes is the possibility to leach small amounts of materials using a process that consumes low energy levels. Maneesuwannarat et al. (2016b) also used a microorganism (*Cellulosimicrobium funkei*) to obtain gallium, achieving high efficiency in the process. The authors discussed the possibility to use the method to recycle semiconductors.

9.7 Critical Elements

The manufacture of electrical and electronic devices depends on the availability of raw materials that are considered geologically critical elements due to the fact that these resources are increasingly scarce in the natural environment. For this reason, these elements are known as scarce metals and rare earth elements, or critical elements. These elements are used mainly in the production of permanent magnets, bulbs, rechargeable batteries, catalysers, and other devices (Binnemans et al. 2013). In most cases, these metals replace materials with the same function; alternatively, they are added to improve the performance of devices. Most devices associated with energy efficiency – like LEDs – utilize geologically critical elements. Obtaining these elements is, therefore, essential in the manufacture of these products.

Rare earth elements, also called lanthanides, are represented by Ln and include 15 elements on the sixth period of the periodic table, between elements 57 and 71 (lanthanum (La), cerium (Ce), praseodymium (Pr), neodymium (Nd), promethium (Pm), samarium (Sm), europium (Eu), gadolinium (Gd), terbium (Tb), dysprosium (Dy), holmium (Ho), erbium (Er), thulium (Tm), ytterbium (Yb), and lutetium (Lu)), besides yttrium and scandium (Serra 2011; Martins and Isolani 2005; Viera and Lins 1997). These elements are present at low levels, although cerium is the most abundant and thulium is the rarest. The average per cent level of rare earth elements on the earth's crust is approximately 0.01%, and over 250 ores are known to contain low levels of these elements. Rare earth elements are divided into two groups: the cerium group (light rare earth elements) and the yttrium group (heavy rare earth elements). This classification was based on their different chemical properties (Viera and Lins 1997; Serra 2011). In addition to these elements, gallium and yttrium are considered critical in the production of electrical and electronic devices. Therefore, recycling is an interesting alternative both from the environmental and economic standpoints.

According to Jones (2013), the value of rare earth elements increased 750% in 2011, warning the technology industry of the need for alternatives to these elements that do not depend on the Chinese market. One such alternative is the recycling of electrical and electronic equipment, which contains a large number of elements that can be recovered. Recycling aims to solve the environmental problems caused by the inappropriate disposal of products; however, today recycling plays a relevant role in the supply of scarce materials. It is a new method to procure raw materials, which is called urban mining. Currently, 49 million tons of Electronic-waste is generated every year, and only 10% of that is recycled. In turn, the amount of rare earth elements recycled is only 1% (Jones 2013).

The critical elements used in the LED industry are the metals indium and gallium and the rare earth elements europium and terbium. Serra et al. (2015) claim that the development of lighting technologies is closely linked with the availability of rare earth elements since these play an essential role in semiconductors.

However, the recycling of LEDs faces hurdles that have to be overcome. Valuable metals like indium and gallium and rare earth elements like europium and terbium are found in the inner components of LEDs. Therefore, to recycle them, it is first necessary to remove the materials used for encapsulation like glass, plastic, ceramics, aluminium, and copper resistors. Gallium and indium are the main critical elements recovered from LEDs, while rare earth elements are not extensively recycled due to the lack of proper techniques for the purpose. As shown above, techniques like acid leaching, bacterial leaching, and pyrometallurgical methods may be used to recover these elements, but the concentration of these elements remains the main challenge faced by the industry today (Swain et al. 2015a, b, c; Zhan et al. 2015; Maneesuwannarat et al. 2016a, b). Another difficulty lies in increasing the amount of recycled material to boost production scale, which is problematic in view of the incipient character of waste management in many countries.

Concerning production and price of materials, gallium and indium production values are 216 t/year and 640 t/year, respectively, with prices at US$517/kg and US$561/kg, in that order. The main rare earth elements used in the production of LEDs are europium, lutetium, yttrium, and cerium, all of which are used in light conversion. Annual production values of cerium, europium, lutetium, and yttrium are 24,000 t, 10 t, 10 t, and 8900 t, respectively, while prices are US$36/kg, US$418/kg, US$800/kg, and US$7/kg, in that order. In addition, gold, silver, and tin are also considered critical elements and are widely used in electrical, thermal, and mechanical connections (Fraunhofer Institute 2018).

9.8 Conclusion

The growth of the LED industry shows promise. From the technological standpoint, it is associated with the search for alternatives for the procurement of critical elements like gallium, indium, and rare earth elements. One of these alternatives is recycling, which is an important technique when it comes to protecting the environment and recovering critical elements for the production of new LEDs. Several techniques are being investigated, like pyrometallurgy and hydrometallurgy, but the main challenge is to obtain larger amounts of these elements and to increase scale in a recycling production chain of critical elements that may be returned to the production line of new LEDs.

References

Ascurra RE (2013) Eficiência Elétrica em Iluminação Pública Utilizando Tecnologia Led: um Estudo de Caso. Dissertation, Universidade Federal de Mato Grosso, Cuiabá, MT, Brazil

Associação Brasileira da Indústria de Iluminação (ABILUX) (2017) Guia LED, 5ª edn. http://www.abilux.com.br/portal/pdf/guia_led_5ed.pdf. Accessed 10 June 2017

Ayres RU, Pieró LT (2017) Material efficiency: rare and critical metals. The Royal Society Publishing, Fontainebleau

Bastos FC (2011) Análise da Política de Banimento de Lâmpadas Incandescentes do Mercado Brasileiro. Dissertation, Universidade Federal do Rio de Janeiro, Rio de Janeiro, RJ, Brazil

Binnemans K, Jones PT, Blanpain B, Gerven TV, Yang Y, Walton A, Buchert M (2013) Recycling of rare earths: a critical review. J Clean Prod 51:1–22

Buchert M, Schuler D, Bleher D (2009) Critical metals for future sustainable – technologies and their recycling potential. UNEP, Paris

Bullough JD (2003) Lighting answers: LED lighting systems. Lighting Research Center. v. 7, http://www.lrc.rpi.edu/programs/nlpip/lightinganswers/led/whatisanled.asp. Accessed 15 Feb 2018

Callister WD, Rethwisch DG (2013) Ciência e engenharia de materiais: uma introdução, 8th edn. LTC, Rio de Janeiro

Carvalho TCMB, Xavier LH (2014) Gestão de Resíduos Eletroeletrônicos: Uma Abordagem Prática para a Sustentabilidade, 1st edn. Elsevier, XVII, Rio de Janeiro

Castro DB (2013) Iluminação por LEDs. Especialize. Revista Especialize On-line IPOG. 5(1). https://www.ipog.edu.br/revista-especialize-online/edicao-n5-2013/iluminacao-por-leds. Accessed 10 Feb 2018

CEMPRE (Compromisso Empresarial em Reciclagem) (2018) Lâmpadas de LED a Caminho da Reciclagem Economicamente Viável. http://cempre.org.br/informa-mais/id/51/lampadas-de-led-a-caminho-da-reciclagem-economicamente-viavel. Accessed 6 Feb 2018

Cervi M (2005) Rede de Iluminação Semicondutora para Aplicação Automotiva. Dissertation, Universidade Federal de Santa Maria (UFSM), Santa Maria, RS, Brazil

Cycled (2017) End of life. http://www.cyc-led.eu/End%20of%20life.html. Accessed 27 May 2017

Denbaars SP (1997) Gallium-nitride-based materials for blue to ultraviolet optoelectronics devices. Proc IEEE 85(11):1740–1749

Dias MP (2012) Avaliação do Emprego de Um Pré-Regulador Boostde Baixa Frequência do Acionamento de Leds de Iluminação. Dissertion, Universidade Federal de Juiz de Fora, Juíz de Fora, MG, Brazil

Directive 2012/19/UE of the European Parliament and of the Council of 4 July 2012 on waste electrical and electronic equipment (WEEE). Official J L 197, 59. http://eur-lex.europa.eu/legal-content/PT/TXT/PDF/?uri=OJ:L:2012:197:FULL&from=PT. Accessed 12 Aug 2017

Fecomércio SP (2016) COMÉRCIO registra alta de mais de 100% nas vendas de lâmpadas de LED. http://www.fecomercio.com.br/noticia/comercio-registra-alta-de-mais-de-100-nas-vendas-de-lampadas-de-led. Accessed 14 Oct 2017

Fraunhofer Institute (2018) https://www.izm.fraunhofer.de/de/abteilungen/environmental_reliabilityengineering/projekte/cycled.html. Accessed 5 Feb 2018

Gassmann A, Zimmermann J, Gauß R, Stauber R Gutfleisch O (2016) Led lamps recycling technology for a circular economy. Latest LPR Megazine. LpR Article, N. Aug 25, 2016. https://www.led-professional.com/resources-1/articles/led-lampsrecycling-technologyfor-a-circular-economy. Accessed 15 July 2018

Gois A (2008) LEDs na Iluminação Arquitetural Lighting Now Ed 1. http://www.ebah.com.br/content/ABAAAe77YAD/livro-leds-na-iluminacao-arquitetural?. Accessed 15 July 2018

Jang D, Yook S, Lee K (2014) Optimum design of a radial heat sink with a fin-height profile for high-power LED lighting applications. Appl Energy 116:260–268

Jones N (2013) A escassez de alguns metais raros é um obstáculo para as tecnologias verdes. Yale Environ 360. http://e360yale.universia.net/a-escassez-de-alguns-metais-raros-e-um-obstaculo-para-as-tecnologias-verdes/?lang=pt-br. Accessed 7 Dec 2017

Kiddee P, Naidu R, Wong MH (2013) Electronic waste management approaches: an overview. Waste Manag 33:1237–1250

Kitai A (2011) Principles of solar cells, LEDs and diodes. The role of the PN junction. Wiley, Mississauga

Maneesuwannarat S, Teamkao P, Vangnai AS, Yamashita M, Thiavetyan P (2016a) Possible mechanism of gallium bioleaching from gallium nitride (GAN) by Arthrobacter creatinolyticus: role of amino acids/peptides/proteins bindings with gallium. Process Saf Environ Prot 103:36–45

Maneesuwannarat S, Vangnai AS, Yamashita M, Thiavetyan P (2016b) Bioleaching of gallium from gallium arsenide by Cellulosimicrobium funkei and its application to semiconductor/electronic wastes. Process Saf Environ Prot 99:80–87

Martins TS, Isolani PC (2005) Terras raras: aplicações industriais e biológicas. Quim Nova 28(1):111–117

OSRAM (2009) Life cycle assessment of illuminants a comparison of light bulbs, compact fluorescent lamps and LED lamps, innovations management. Regensburg, Germany. http://www.energ-etica.eu/mediapool/99/993141/data/OSRAM_LED_LCA_Summary_November_2009.pdf. Accessed 24 July 2017

Polanco SLC (2007) A situação da destinação pós-consumo de lâmpadas de mercúrio no Brasil. Dissertation, Escola de Engenharia Mauá do Centro Universitário do Instituto Mauá de Tecnologia, Brazil

Reuter MA, Van Schaik A (2015) Product-centric simulation-based design for recycling: case of LED lamp recycling. J Sustain Metall 1:4–28

Scholand M J, Dillon HE (2012) Life-cycle assessment of energy and environmental impacts of LED lighting products. Part 2: LED manufacturing and performance. US Department of Energy. http://www.pnnl.gov/main/publications/external/technical_reports/PNNL-21443.pdf. Accessed 17 June 2017

Schubert EF (2006) Light-emitting diodes, 2nd edn. Cambridge University Press, Cambridge

Serra AO (2011) Terras Raras – Brasil x China. J Braz Chem Soc 22(5):809

Serra OA, Lima JF, De Sousa Filho PC (2015) A Luz e as Terras Raras. Rev Virtual Quim 7(1):242–264

Stasiak F (2013) A Física dos LEDs. Revista da Set. n. 136. http://www.set.org.br/revista-da-set/a-fisica-dos-leds/. Accessed 27 May 2017

Swain B, Mishra C, Kang L, Park K, Lee CG, Hong HS, Park J (2015a) Recycling of metal-organic chemical vapor deposition waste of GaN based power device and LED industry by acidic leaching: process optimization and kinetics study. J Power Sources 281:265–271

Swain B, Mishra C, Lee CG, Park K, Lee K (2015b) Valorization of GaN based metal-organic chemical vapor deposition dust a semiconductor power device industry waste through mechanochemical oxidation and leaching: a sustainable green process. Environ Res 140:704–713

Swain B, Mishra C, Kang L, Park K, Lee HHS (2015c) Recycling process for recovery of gallium from GaN na e-waste of LED industry through ball milling, annealing and leaching. Environ Res 138:401–408

Teixeira I, Rivera R, Reiff LO (2016) Iluminação LED: sai Edison, entram Haitz e Moore – benefícios e oportunidades para o país. BNDES Setorial, Rio de Janeiro 43:363–412. https://web.bndes.gov.br/bib/jspui/handle/1408/9576. Accessed 12 Aug 2017

Viera EV, Lins FAF (1997) Concentração de minérios de terras-raras: uma revisão. CETEM/CNPq, Rio de Janeiro

Yamachita A, Haddad J, Dias MVX (2006) Iluminação. In: Marques MCS, Haddad J, Martins ARS (eds) Conservação de energia Eficiência Energética de Equipamentos e Instalações, 3rd edn. FUPAI, Itabujá, pp 213–246

Zhan L, Xia F, Ye Q, Xiang X, Xie B (2015) Novel recycle technology for recovering rare metals (Ga, In) from waste light-emitting diodes. J Hazard Mater 299:388–394

Chapter 10
E-waste Management and the Conservation of Geochemical Scarce Resources

Tamires Augustin da Silveira, Emanuele Caroline Araújo dos Santos, Angéli Viviani Colling, Carlos Alberto Mendes Moraes, and Feliciane Andrade Brehm

Contents

Abstract Electrical and electronic equipment (EEE) generates very complex waste due to the wide variety of components such as metals, polymers, ceramic materials, and composite elements. In addition, the growing consumption of these devices due to technological development increases the rate they are disposed of. When improp-

T. A. da Silveira · E. C. A. dos Santos · A. V. Colling
C. A. M. Moraes (✉) · F. A. Brehm
Universidade do Vale do Rio dos Sinos – UNISINOS, Porto Alegre, RS, Brazil
e-mail: cmoraes@unisinos.br

A. Khan et al. (eds.), *E-waste Recycling and Management*,
Environmental Chemistry for a Sustainable World 33,
https://doi.org/10.1007/978-3-030-14184-4_10

erly disposed of, waste electric and electronic equipment (WEEE) may trigger environmental impacts and negative effects on health. Also, the expansion of the electronic industry is based on the extraction of natural resources, some of which are running increasingly scarce. In this scenario, recycling stands as an alternative in the effort to recover economically interesting materials such as metals, which are abundant in waste electric and electronic equipment. This text discusses the current scenario in the electrical and electronic equipment industry and generation of waste electric and electronic equipment considering the implications of resource management and environment, social, and economic impact in this production chain.

Keywords E-waste · Scarce resource · Economic, environmental, and social issues · Recycle process · Urban mining · Elements recovery · Reclamation · Rare earth elements · Critical metals · Hitchhike metal/element

10.1 Introduction

In addition to the reduction in prices of electrical and electronic equipment (EEE), the extraordinary technological development and impressive economic growth of this industry have led to an increase in sales in the past two decades. At the same time, the life cycle of electrical and electronic equipment was reduced dramatically, leading to a significant increase in end-of-life electronic devices, commonly known as waste electric and electronic equipment (WEEE). Current estimates say that roughly 45 million tons of waste electric and electronic equipment are generated every year in the world, and it is said that this number is growing exponentially (Ghosh et al. 2015).

The electronic industry is characterized by technological dynamism (ABINEE 2009). The fast pace of technological development continually presents new, state-of-the-art cell phones, inducing customers to replace devices more often (Sarath et al. 2015). Obsolescence of electrical and electronic equipment is a result of factors like innovative design and new functionalities such as high computing power and speed to carry out tasks, in addition to unrestrained consumption. The useful life of these devices is increasingly short, which translates to higher amounts of waste and consumption of natural resources (Sena 2012).

Compared with conventional kinds of waste, waste electric and electronic equipment pose singular, complex challenges such as the variety and evolution of products in terms of size, weight, functions, and materials used in manufacture. The constant introduction of new products also requires the development of new end-of-life treatment approaches. Another important aspect is the fact that electrical and electronic equipment are produced using high amounts of metals like gold, silver, indium, and platinum, as well as rare earth elements. The low availability of these metals in natural environments and the technological complexity associated with the recycling of these elements are other important variables in waste electric and electronic equipment management (GSMA 2015).

Waste electric and electronic equipment are chemically and physically different from solid waste in general, such as municipal and industrial refuse. Although this kind of waste contains valuable metals, it also contains hazardous materials that require special treatment and recycling methods to prevent environmental contamination and health hazards. Recycling methods afford to recover several components, like copper and precious metals. Nevertheless, due to the lack of proper facilities, high operation costs, and stringent regulations, waste electric and electronic equipment is not recycled in many nations. Instead, this kind of waste is disposed of in common or special landfills or even exported from wealthy nations to underdeveloped ones, where they may be recycled using methods that suffer from poor recovery rates and that do not follow labor safety or environmental protection guidelines (Cobbing 2008).

Even though such high amounts of waste pose challenges to any management system, waste electric and electronic equipment still represent a secondary source of valuable resources in that these are more easily available for recycling (Parajuly et al. 2017). Several components may be found in electrical and electronic equipment, whether they are metallic, plastic, or other substances. In cell phones, for instance, metals account for approximately 23% of the weight of a typical device. The remaining material consists of plastics and ceramic elements. Interestingly, one cell phone may contain as many as 40 chemical elements (UNEP 2009).

The main hazardous elements in waste electric and electronic equipment include lead, cadmium, mercury, and chromium (VI), in addition to polybrominated compounds like polybrominated biphenyls and polybrominated diphenyl ethers (PBDE), which are used as flame retardants (Gouveia 2014).

Cell phones are manufactured using valuable natural resources and materials like polymers, glass, and especially metals. The metal used is obtained by extraction from ores, in energy-intensive processes. Recycling these devices at the end of their useful life represents the opportunity to save natural resources since it avoids the need to extract ores. Another advantage is the reduction of soil and water pollution caused when they are improperly disposed of. Also, air pollution is mitigated due to the lower emissions of greenhouse gases inherent to the processes used to extract new raw materials.

This chapter discusses three main issues: (i) the growth of the international electrical and electronic equipment market and the impact thereof on the demand and supply of elements used in the manufacture of these devices, (ii) the global scenario of waste electric and electronic equipment, and (iii) the technology for the management and the conservation of geochemically scarce resources. The topics are presented using basic concepts such as the definition and classification of electrical and electronic equipment and include a more in-depth discussion of the composition of these devices and the environment and social impacts associated with the extraction of resources used to produce them and the end-of-life fate of waste electric and electronic equipment.

10.2 Electrical and Electronic Equipment

10.2.1 Definition

One of the statutory definitions of electrical and electronic equipment (EEE) was provided by the European Union, which succinctly describes it as equipment that uses electric current or electromagnetic field and equipment used in processes such as generation, transfer, and measurement of currents and fields. This definition is based on the categories defined in Annex I, Directive 2012/19/EU, which covers electrical and electronic equipment of various sizes and includes specific groups (such as medical devices) and divides electrical and electronic equipment into ten categories (Box 10.1).

Box 10.1 Electrical and Electronical Equipment Categories According to Directive 2012/19/EU

Category	Electric and electronic equipment of each category
Large household appliances	Large cooling appliances, refrigerators, freezers, other large appliances used for refrigeration, conservation, and storage of food; washing machines, clothe dryers, dishwashing machines; cookers, electric stoves, electric hot plates, microwaves, other large appliances used for cooking and other processing of food; electric heating appliances, electric radiators, other large appliances for heating rooms, beds, seating furniture; electric fans, air conditioner appliances, other fanning, exhaust ventilation, and conditioning equipment
Small household appliances	Vacuum cleaner; carpet sweepers; other appliances for cleaning; appliances used for sewing, knitting, weaving, and other processing for textiles; irons and other appliances for ironing, mangling, and other care of clothing; toasters; fryers; grinders, coffee machines, and equipment for opening or sealing containers or packages; electric knives; appliances for hair cutting, hair drying, tooth brushing, shaving, massage and other body care appliances; clocks, watches, and equipment for the purpose of measuring, indicating, or registering time; scales
IT and telecommunications equipment	Mainframes; minicomputers; printed units; personal computers (CPU, mouse, screen, and keyboards included); laptop computers; notebook computers; notepad computers; printers; copying equipment; electrical and electronic typewriters; pocket and desk calculators; other products and equipment for the collection, storage, processing, presentation, or communication of information by electronic means; user terminals and systems; facsimile machine (fax); telex; telephones; pay telephones; cordless telephones; cellular telephones; answering systems; other products or equipment of transmitting sound, images, or other information by telecommunications

(continued)

Box 10.1 (continued)

Category	Electric and electronic equipment of each category
Consumer equipment and photovoltaic panels	Radio sets, television sets, video cameras, video recorders, audio amplifiers, musical instruments, other products or equipment for the purpose of recording or reproducing sound or images, including signals or other technologies for the distribution of sound and image than by telecommunications
Lighting equipment	Luminaires for fluorescent lamps with the exception of luminaires in households; straight fluorescent lamps; compact fluorescent lamps; high-intensity discharge lamps, including pressure sodium lamps and metal halide lamps; low-pressure sodium lamps; other lighting or equipment for the purpose of spreading or controlling light with the exception of filament bulbs
Electrical and electronic tools (except large-scale stationary industrial tools)	Drills; saws; sewing machines; equipment for turning, milling, sanding, grinding, sawing, cutting, shearing, drilling, making holes, punching, folding, bending or similar processing of wood, metal, and other materials; tools for riveting, nailing or screwing or removing rivets, nails, screws or similar uses; tools for welding, soldering, or similar use; equipment for spraying, spreading, dispersing, or other treatment of liquid or gaseous substances by other means; tools for mowing or other gardening activities
Toys, leisure, and sports equipment	Electric trains or car racing sets; handheld video game consoles; video games; computers for biking, diving, running, rowing; sports equipment with electric or electronic components; coin slot machines
Medical devices (except all implanted and infected products)	Radiotherapy equipment, cardiology equipment, dialysis equipment, pulmonary ventilators, nuclear medicine equipment, laboratory equipment for in vitro diagnosis, analyzers, freezers, fertilization tests, other appliances for detecting, preventing, monitoring, treating, alleviating illness, injury, or disability
Monitoring and control instruments	Smoke detector; heating regulators; thermostats; measuring, weighing, or adjusting appliances for household or as laboratory equipment; other monitoring and control instruments used in industrial installations (e.g., in control panels)
Automatic dispensers	Automatic dispensers for hot drinks, automatic dispensers for hot or cold bottles or cans, automatic dispensers for solid products, automatic dispensers for money, all appliances which deliver automatically all kinds of products

In Brazil, the Brazilian Industrial Development Agency (ABDI 2013) classifies electrical and electronic equipment into four large groups that are represented by colors, namely, green, brown, white, and blue. For example, green goods are small, have a short life cycle (between 2 and 5 years), and include notebooks and cell phones. Figure 10.1 illustrates the types of electrical and electronic equipment included in each line and respective life cycles.

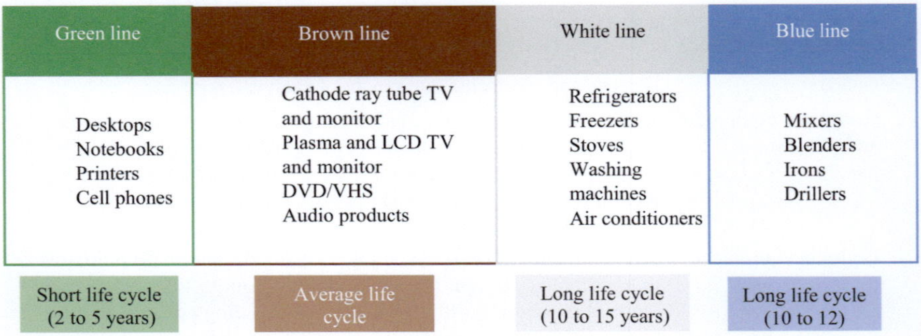

Fig. 10.1 Life cycle and categories of electrical and electronic equipment according to the classification established by the Brazilian Industrial Development Agency. (*Source*: Adapted from ABDI (2013))

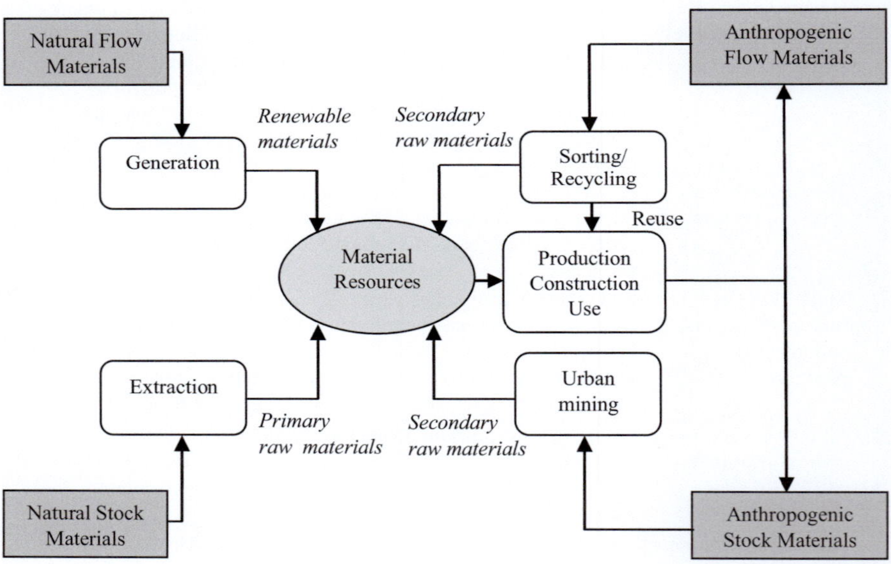

Fig. 10.2 Description of material flows between sources of material resources. (*Source*: Cossu and Williams 2015)

For the US Environmental Protection Agency (EPA), electrical and electronic equipment are devices that run on electrical currents or electromagnetic fields. These devices include the appliances used in the generation, transfer, and measurement of these currents and fields. In other words, if a product requires a battery or a power supply, it is considered an electrical and electronic equipment. Moreover, this definition establishes that the design of these devices should address reuse or recycling at the end of the life cycle (EPA 2018) (Fig. 10.2).

10.2.2 The International Market of Electrical and Electronic Equipment

In the past two decades, the electronic industry and the communication and information technology sector, in particular, revolutionized the world. Today, electrical and electronic equipment is ubiquitously present in homes, offices, and organizations across the globe. So it comes as no surprise that life in both developed and developing nations would be simply possible without these devices (GSMA 2015).

Research has shown that the international market of electrical and electronic equipment like TVs, refrigerators, washing machines, and sound systems has reached a plateau in developed countries. But these markets continue to expand in developing nations due to perceptibly apparent factors like population growth and the increase in the income of families. This scenario explains the growth in the generation of waste electric and electronic equipment (WEEE) per person in Latin America, compared with the rest of the world. For Araújo (2013), electrical and electronical equipment markets in developed countries have reached a full maturity status. Yet, a different trend is observed for computers, tablets, and cell phones, whose sales in general continue to increase worldwide.

Besides the fast-paced technological evolution and the search for high-quality electrical and electronic equipment and services accessed through these devices, improved power of purchase induces consumers to replace electrical and electronic equipment more often today compared to what was seen in the recent past (Kasper 2011). Today it is also clear that despite market saturation, the fast replacement of technologies signals continually growing sales, given that consumers will naturally wish to buy new, more technologically advanced goods (Araújo 2013).

10.2.3 Supply and Demand of Chemical Elements in the Manufacture of Electrical and Electronic Equipment

Technological advances and the increasing demand for state-of-the-art devices are the main factors behind the high requirements for chemical elements in the production of electrical and electronic equipment in the near future. For Henckens et al. (2016), the volumes of ore extracted from the Earth increase sharply at the same time that sources dwindle; such is the case of elements like antimony, molybdenum, and zinc (Zn), all of which are essential in the production of electrical and electronic equipment. The world faces the actual risk of running out of these elements within a century or even a few decades if volumes extracted continue to increase at current rates.

Sustainable industrial growth depends on stable supplies of natural resources. Metals are increasingly important in the expansion of industrial sectors like information technology, transportation, and steel production. Nevertheless, the growth of a given industrial sector depends not only on the efficient, sustainable procurement of domestic resources but also on the critical analysis of resource availability in international markets. In this scenario, urban mining emerges an interesting alternative in the effort to obtain scarce metals from electrical and electronic equipment at the end of their life cycle (Jung Jo et al. 2017).

In addition to economic incentives, measures like effective environmental policies, methods to improve environmental awareness, the placement of recycling banks in suitable locations, and the development of efficient mechanisms to collect electrical and electronic equipment at the end of their life cycle should be devised (Tesfaye et al. 2017). Improved separation, collection, and recycling of electrical and electronic equipment are now seen as effective ways of obtaining metals, contributing to a sustainable economy and environment conservation and reducing the demand for natural resources.

The geological scarcity of mineral resources has to be distinguished from the economic shortage. The latter is a consequence of the former, with many associated factors at play. The main difference between economic and geologic scarcity is the fact that the latter is a structural and physical phenomenon, while the former is more cyclical in character (Henckens et al. 2016). Market price is determined by the balance between supply and demand. Demand for mineral resources increases as a result of the development of new applications, as observed in the use of rare earth elements (REE) in electrical and electronic equipment, for instance. But the fast industrial development in large countries such as China is another factor that increases the demand for ores. In turn, demand may fall in the wake of a new, cheaper replacement. Supply may also be influenced by decisions made by an oligopoly or monopoly. On the other hand, factors that reduce supply include accidents, strikes, and geopolitical actions.

Wen et al. (2015) developed a model to predict the demand, recycling potential, and availability of copper (Cu), aluminum (Al), lead (Pb), and iron (Fe). These metals are highly consumed in China, which was the largest steel producer in 1967, 1980, 1990, 2000, and in every year between 2007 and 2017, the year when the country produced 49.7% of the world's steel (Wen et al. 2015; World Steel Association 1978, 2010, 2011, 2012, 2013, 2014). In 2015 China led the production of pig iron and was the main producer of Al between 2010 and 2017 (USGS 2016; Statista 2017). Research indicates that primary copper and iron resources in China will be exhausted in 10 and 30 years, respectively. This means that mining metals will be a difficult task in the future. However, there are alternative ways to secure a stable supply of metals, based on the recovery of metals in electrical and electronic equipment, for instance.

The primary copper and iron resources may be effectively replaced by secondary metals, as shown in Table 10.1.

It has been estimated that primary resources will be replaced by other sources in the years between 2020 and 2040. These other sources include the recycling and recovery of metals like copper and iron in E-waste, for instance, whose recovery

Table 10.1 Indicators of Cu, Fe, Al, and Pb

Resource name	Year	Domestic exploration	Import	Scraps	Recycling	Substitution rate (%)
Copper (10,000 tons)	2020	97.9	525.5	417.2	330.5	34.6
	2030	70.9	608.8	778.8	607.8	47.2
	2040	51.4	631.7	1050.1	808.7	54.2
Iron (million tons)	2020	223.9	438.8	257.1	223.5	25.2
	2030	156.7	295.4	538.7	471.7	51.1
	2040	0	248.3	808.8	711.6	74.1
Aluminum (10,000 tons)	2020	949.0	3441.1	2337.9	930.4	17.5
	2030	803.3	5374.9	6269.9	2472.6	28.6
	2040	680.0	6603.5	9388.4	3697.9	33.7
Lead (10,000 tons)	2020	0	678.4	101.85	549.3	44.7
	2030	0	1029.5	1860.9	1012.9	49.6
	2040	0	1130.6	2219.3	1208.2	51.7

Note: Substitution rate = recycling/(domestic exploration + import + recycling)
Source: Modified from Wen et al. (2015)

rates will increase by 19.6% and 48.8%, respectively. In addition, copper, aluminum, and lead import and export figures are also expected to rise.

Rare earth elements are also widely used in the electronic industry. These metals have exclusive properties and are used as doping agents in semiconductors, printed circuit boards (PCB), catalyzers, and photovoltaic cells, among other applications (Ayres and Peiró 2013; Loureiro 2013).

In 2010 the European Commission looked into the availability of 41 raw materials and discovered that the supply of rare earth elements was at the highest risk of shortage. In 2014 the list was reviewed and the same conclusion was reached. This is due to the abundance of rare earth elements on the Earth's crust, compared with other metals, in addition to other challenges associated with the extraction of rare earth elements in the environment as well as processing and global availability, like:

- Geological distribution: the ores containing rare earth elements rarely occur as concentrated forms or in individually, which makes exploration more difficult.
- Mining risks: the elements usually occur next to uranium (U) or thorium (Th) decay chains, which makes extraction even more complex due to the radiotoxicity hazard.
- Environmental process and impact: mining, leaching, pre-concentration, and the various stages required to reach purity degrees needed in certain applications, which may be as high as 99.99%, normally generate large amounts of waste and effluents (European Commission 2010, 2014; Tunsu et al. 2015).

China produces 97% of the 125,000 tons of rare earth elements in the world, and there is a strong demand for new ore deposits to extract these elements (Loureiro 2013). Nevertheless, urban mining is a potential alternative to mitigate the scarcity of resources in that country.

10.2.4 Economic, Environmental, and Social Issues

In general, mining has several negative environmental, economic, and social impacts associated with the extraction and processing of ores. The removal of plant covers is one of the first stages of ore extraction, when environmental impacts start, followed by subsequent impacts linked with operations, such as the production of dust, generation of a large amount of solid waste (tailings), and liquid effluents. For example, data published by Tunsu et al. (2015) indicate that each ton of rare earth elements produced generated approximately 8.5 kg of fluoride (F) and 13 kg of dust. In order to obtain concentrated rare earth elements material, 75 m^3 of acid effluents and between 9.6 m^3 and 12 m^3 of gas are produced. This gas includes sulfuric acid and sulfur dioxide. Approximately 1 ton of radioactive waste is also generated.

Several mineral reserves are located in forests, in areas that are rich in biodiversity, and regions inhabited by native populations. Environmental degradation and the replacement or even the loss of subsistence means due to the expansion of mining activities have caused serious conflicts with local populations and society at large (Arora et al. 2017).

In developing countries, tailings generated by mining are disposed of in basins, many times with no control, causing accidents such as the one that occurred in Mariana, Brazil. In November 2015, basins holding tailings burst due to excess volume, releasing 34 million m^3 of iron tailings in rivers and lakes, killing 17 people, polluting rivers and lakes, killing animals and plants, and causing impacts that have not been fully assessed as yet.

The mining industry has a tragic record in health, human safety, and environmental issues. Environmental monitoring is compulsory in most plants. In many cases, the adoption of sustainable practices is imposed by regulation or initiatives from nongovernmental organizations. These changes affect mining and processing in operations of foundries and refineries. For example, refineries have taken the required steps to reduce the emission of effluents (Arndt et al. 2017). However, solid waste, including tailings and scum from foundries and refineries, still pose significant hazard concerning the safe storage of large amounts of toxic elements and potential release of these effluents (Arndt et al. 2017).

10.3 Waste Electronic and Electrical Equipment

10.3.1 What Is E-waste?

The literature lists several ways to refer to this kind of waste, including the acronym waste electric and electronic equipment. However, the most accepted definition worldwide was given by Directive 2012/19/EU, according to which E-waste is the waste generated from electrical and electronic equipment after the end of their life cycle and includes all components, subsets, and usable materials that are part of the device when it is disposed of (União Europeia 2012).

For Shagun et al. (2013), E-waste is a popular name for electrical and electronic equipment at the end of their life cycle. The term is possibly the most used in technical literature and encompasses several forms of electrical and electronic equipment that have lost their value to their owners (Garlapati 2016).

The Brazilian Centro de Tecnologia Mineral calls electronic or technological waste any waste generated from the disposal of electronic equipment that may be recycled, like computers, cell phones, portable electronics, TVs, batteries, radios, and other devices (CETEM 2015).

For Nicolai (2016), E-waste is composed of the raw materials obtained from waste generated after the use of electronic equipment that is not usable anymore and is directed to recovery, recycling, or disposal.

E-waste is generated when electrical and electronic equipment has lost their value or cannot be used as originally designed. This happens because the electrical and electronic equipment are replaced by new, more advanced products or are damaged beyond repair. The increasing array of electrical and electronic equipment that ends as E-waste includes home appliances like refrigerators, air conditioning systems, cell phones, and computers (Jaiswal et al. 2015; Britannica Academic 2016).

Currently, E-waste is considered the most widely generated waste in the world, with growth rates at between 3% and 5% a year (Cucchiella et al. 2015).

10.3.2 The Global Scenario of E-waste Generation

It has been estimated that 41.8 million tons of E-waste were generated in 2014, representing 5.9 kg per person, at the global level (GSMA 2015; STEP 2015). The largest amounts of E-waste were generated from small (13 million tons) and large (11 million tons) devices (Table 10.2). As a rule, E-waste is classified into five or six categories, from small devices to a specific group that includes screens.

Data from GSMA (2015) indicate that most E-waste is generated in Asia, which contributed 16 million tons of this kind of waste in 2014 (38% of the total amount of this waste in the planet). Europe comes second, generating 28% of the world's E-waste, followed by North America (19%), Latin America (9%), Africa (5%), and Oceania (1%) (GSMA 2015). According to a report published by the United Nations Environment Program, the electronic sector generates 41 million tons of E-waste every year, and it has been estimated that this value would reach 50 million tons in 2017 (UNEP 2015). The report also informed that between 60% and 90% of this waste is illegally sold or disposed of as common waste.

10.3.2.1 Impact of Waste Generation at the Global Level and Flow

The illegal transportation of hazardous waste from developed countries to developing nations causes great concern worldwide due to high treatment and disposal costs and ineffective environmental regulations (or absence thereof), in addition to poor

Table 10.2 Amount of E-waste generated from various categories of electrical and electronic equipment in 2014 in the world

Flow/category	Equipment	Quantity (million tons)
Small devices	Vacuum cleaners, microwave ovens, hair and body care devices, recorders, radios	12.8
Large devices	Washing machines, stoves, dishwashers	11,8
Temperature control equipment	Refrigerators, freezers, air conditioners, heat pumps	7.0
Screens	Screens, monitors, and TV, netbooks and notebook screens	6.3
Small telecommunication equipment and devices	Cell phones, GPS devices	3.0
Bulbs	Any kind of bulb	1.0

Source: Prepared by the authors based on GSMA (2015)

environmental awareness (UNEP 2015). A study carried out by the USEPA revealed that it is ten times cheaper to export E-waste to Asia than process it in the USA (Lundgren Lundgreen 2012).

The management and fate of E-waste are approached differently in developing and developed nations. In numerous cases, the main approach to dispose of E-waste is the straightforward dumping in common landfills. The potential environmental hazards caused by E-waste include the contamination of soil and water bodies due to the increased mobility of metals and organic compounds in landfills, even after these are closed (Tansel 2017).

The exact dimension of the illegal E-waste trade worldwide is not known (UNEP 2015). As a rule, E-waste flows to historically disempowered, low-income populations. According to the UNEP (2015), recycling of E-waste is a flourishing business in several parts of the world but mainly in Southeast Asia, India, and Pakistan. More specifically in West Africa, the main importers of E-waste are Ghana, Nigeria, and Benin.

For Lundgreen (2012), the main substances that may be released during the recovery and recycling of E-waste fall within three main groups: (i) original components of the equipment (lead and mercury), (ii) substances that may be added during recovery processes (cyanide), and (iii) substances that may form during recycling (dioxins, which may be released during the burning of plastics in computer housings and cables). If E-waste management is not carried out properly, such substances may pose significant hazards to the environment and human health. Lundgreen (2012) claims that toxic substances may also be released during informal E-waste recycling processes, including leaching, physical disassembly (which generates particulate material and effluents, some of which may contain cyanide), burning (which generates ash), and heating (which releases mercury during desoldering).

Risks to human health include difficulty to breathe, respiratory irritation, cough, asphyxia, pneumonitis, tremors, neurological problems, convulsion, coma, and even death (Yu et al. 2006).

Also, Song and Li (2014) assessed how metals in E-waste cause environmental impact in China. The study revealed the existence of soil, air, water, and plant contamination as well as the accumulation of metals in crops like rice. The concentrations of copper (Cu), mercury (Hg), chromium (Cr), and lead (Pb) were extremely high, compared with the maximum permitted values.

10.3.2.2 The Elements Present in Electrical and Electronic Equipment

Umicore (2016) considers printed circuit boards (PCB) as deposits of metals suitable for urban mining since the concentration of metals they contain is higher than the values obtained through primary mining. For instance, recycling of 1 ton of computer boards produces 250 g of gold (Au).

For the sake of comparison, 1 ton of ore contains 1.5 g of gold, while the amount of the metal in 1 ton of printed circuit boards and 1 ton of cell phones is between 150 g and 200 g and 250 g and 300 g, respectively. For Nicolai (2016), it would be necessary to process 60 tons of gold ore to obtain the equivalent amounts of the metal. In turn, copper levels obtained from direct mining are below 1%, while the amounts of the metal in printed circuit boards are between 15% and 18%, even though a recent study showed that it is possible to obtain as much as 22% of copper from this component (da Silveira et al. 2016; Umicore 2016).

For Wen et al. (2015), primary sources of copper and iron may be effectively replaced by secondary sources of the metals. It has been estimated that replacement rates for copper and iron will increase by 19.6% and 48.8%, respectively, between 2020 and 2040. The authors also claim that the potential exploration of primary resources in China will continue to grow in the same period, despite the increasing importance of recycling E-waste as a source of metals from secondary sources for the country's economy.

Research shows that common metals such as copper and precious metals like rare earth elements may be extracted from E-waste (Palmiere et al. 2014). Rare earth elements are difficult to extract and create mining hazards due to the presence of radioactive elements such as uranium (U) in ores, especially when extraction methods are not carried out properly (Rare Element Resources 2016).

Also, each ton of rare earth elements generates approximately 8.5 kg of fluorine and 13 kg of dust. The production of concentrated rare earth elements generates 75 kg of acid effluents, between 9600 m^3 and 12,000 m^3 of gas containing sulfuric acid and sulfur dioxide, in addition to 1 ton of radioactive waste (Tunsu et al. 2015). Moreover, rare earth elements sometimes are present in continuous ore bodies, which means that it is necessary to mine large areas to obtain these elements (Tunsu et al. 2015). Ayres and Peiró (2013) claim that rare earth elements are never found at high concentrations in the environment. On the contrary, they occur as contaminants or trace elements in ores of elements called attractor metals, to which they are chemically similar. In rare earth elements, the attractor metal is iron. Other rare metals are found in copper (Cu), lead (Pb), nickel (Ni), or zinc (Zn) ores. Attractor metals are mined in large amounts, like iron (Fe), aluminum (Al), copper (Cu), nickel (Ni), zinc (Zn), and others, while hitchhiker metals are geologically rare and chemically related with one of the attractor elements.

Due to their exclusive properties, rare earth elements are used in electronics, especially as doping agents in semiconductors, printed circuit boards, catalyzers, photovoltaic cells, and other applications, which lists hitchhiker metals and respective applications (Ayres and Peiró 2013; Loureiro 2013). China is responsible for 97% of the 125,000 tons of rare earth elements produced every year in the world, and new deposits are constantly sought (Loureiro 2013). Nevertheless, recent studies have discussed the potential of urban mining to mitigate the scarcity of resources in the country.

10.3.2.3 Recycling and Urban Mining

Urban mining is a new approach to the flow of materials, which allows recovering compounds from various anthropogenic processes (Baccini and Brunner 2012; Lederer et al. 2014). Recovered materials present concentrations of valuable components that are similar to the values observed in the natural environment and may represent a significant supply of resources (Cossu and William 2015).

For Krook and Bass (2013), "urban" means within the boundaries of a city environment, while "mining" is understood as the extraction of secondary metal resources from the obsolete material. The concept of urban mining includes all activities and processes used to recover compounds, energy, and elements from manufactured goods, buildings, and waste generated from urban catabolism (Di Maria et al. 2013; Wen et al. 2015).

According to the US Environmental Protection Agency (USEPA), recycling involves the minimization of solid waste based on the recovery and reprocessing of products that would otherwise be disposed of such as aluminum cans, paper, and bottles. Another definition says that recycling is the process of transforming solid waste based on changing its physical, physicochemical, and biological properties with a view to the generation of production materials or conversion into new products (Brasil 2010).

Currently, the terms "recycling," "resource recovery," "urban mining," "waste minimization," "circular economy," and "material recovery" are being increasingly used in business and industrial environments (Cossu and William 2015). Independently of terminology choices, recycling is important not only from the economic aspect but also from concern with the increasing scarcity of primary raw materials in the effort to ensure supplies to future generations. The flow of materials based on extraction from natural environments is finite, while recycling and urban mining will be the most common approaches in the future.

The use of urban mining to obtain rare earth elements is directly associated with the recovery of precious metals based on the recycling of devices. Nicolai (2016) claims that more appropriate and economically feasible methods have to be considered in this form of recycling. For example, Umicore uses a combination of pyrometallurgical, hydrometallurgical, and electrometallurgical processes to recycle waste to recover about 20 precious metals and nonferrous metals (Kasper 2011; Santanilla 2012).

10.3.2.4 Technology for the Management and Conservation of Geochemically Scarce Resources

The topics discussed below address the technologies currently used to recover rare earth elements. If urban mining is to become a feasible alternative, it is extremely important that reverse logistics and recovery technologies are consolidated on a higher scale.

The main processes used to recycle rare earth elements are pretreatment (which consists of the screening and disassembly of devices), mechanical processing, and metallurgical processing. This step is divided into pyrometallurgical and hydrometallurgical processes (Veit 2001; Vivas and Costa 2013; Oliveira 2012).

10.3.2.4.1 Disassembly

Disassembly is the first process used to obtain rare earth elements. It consists of the removal of toxic compounds and the selection of valuable elements that may be separated manually or automatically as a means to simplify the further stages in the recovery of elements (de Moraes 2011). Therefore, the objective of this stage is to optimize the next recycling or recovery stages, whether these are mechanical or metallurgical in nature (de Moraes 2011; Silvas 2014).

For Cui and Forssberg (2003), most recycling organizations use manual disassembly as the first step. For Knoth et al. (2000), disassembly is the first and most important step in recycling and is carried out manually. This makes disassembly more time-consuming and hazardous, increasing the difficulty to perform the process on an industrial scale. The lack of standardized equipment, which is produced by different manufacturers, is another aspect that hinders any move toward perfecting disassembly. Significant improvements in this important step will have to be considered in the next few years. Knoth et al. (2000) also claim that due to increasing amount of products that are recycled, it is necessary to automatize processes as a means to increase efficiency both in technical and economic aspects.

After disassembly, materials are treated according to their characteristics. For Oliveira (2012), the advantages of disassembly and screening include the recovery of components, the removal of potentially hazardous materials, and the segregation of fractions of different added values.

10.3.2.4.2 Mechanical Processing

Mechanical processing, also called physical processing, aims to segregate metals, polymers, and ceramic materials using milling, classification based on grain size, and separation (Gerbase and Oliveira 2012).

For Moraes (2011), the mechanical treatment of printed circuit boards may require milling, grain size classification, magnetic segregation, electrostatic segregation, and heavy medium separation, amid other processes.

Other authors, like Ventura (2014), include disassembly and screening as stages of the mechanical processing of waste. The other stages are milling, grain size classification, magnetic segregation, electrostatic segregation, and gravimetric separation.

Milling is the mechanical fragmentation of rare earth elements with or without the electronic components on the board (Ventura 2014). The operation involves the controlled reduction of the size of the material. Milling is carried out in a vertical or horizontal knife, hammer, or ball mills (Veit 2001). At this stage, the material is milled using the combination of impact, compression, abrasion, and attrition to appropriate particle size (CETEM 2004). According to Richter (2009), the objective of mechanical processing is to redefine the shape and size of particles, increasing surface area to promote chemical reactions and release metals from the part that has no use after ore fragmentation. Milling also influences the segregation of organic from inorganic materials, improving the efficiency of the process. This separation is carried out based on the differences in density or grain size. This stage is carried out using a hammer, knife, or cryogenic mills (Veit 2005). After milling, materials are separated based on grain size or magnetic, electrostatic, and gravimetric methods or even leaching (de Moraes 2011). The classification and sieving of milled material are carried out to obtain two or more factions of particles of different sizes. In sifting, the material is segregated based on particle size, while in classification the material is segregated based on the speed at which particles cross a liquid medium (CETEM 2004). Grain size classification is carried out using shaking screens of different meshes to retain particles. Although this process segregates materials based only on particle size (Hayes 1993; Kasper 2011), Oliveira (2012) claims that components that are attached to boards may be separated by efficient milling.

A given material has its own magnetic characteristics, which depend on a few factors like chemical composition, atom arrangement (electron spin), and crystalline structure, among others (de Moraes 2011). The magnetic properties of materials are also influenced by atomic arrangements and the electronic structure of the elements. Magnetism is influenced by the constituting elements, the concentrations of these elements, and the crystalline structure in the solid (Ribeiro 2013). Materials and minerals are classified based on magnetic susceptibility, which determines the response of a material to a magnetic field. This affords to classify minerals as ferromagnetic, which are attracted by magnetic fields, or paramagnetic, which are only weakly attracted. Materials and minerals repelled by a magnetic field are called diamagnetic (CETEM 2004).

For Oliveira (2012), several models of magnetic separators are widely used, especially dry and wet devices, which are used to separate large and small particles, respectively. Drum, magnetic roll, and cross-belt separators are some of the devices used (CETEM 2004). Electrostatic separators are used to segregate conductor from nonconductor materials, especially to recover Cu and metals from printed circuit boards, in addition to the recovery of aluminum and copper from cables and wires (Cui and Forssberg 2003; Veit 2005). Veit (2005) states that the large difference between the electric conductivity of metals and nonmetal elements establishes an excellent condition to use this technology successfully in the recycling of waste. For the author, the first electrostatic separators were developed to process ores. Today, these devices are used for other aims, like the recovery of nonferrous metals from automobile scrap, the treatment of municipal solid waste, the processing of electrical and electronic equipment, and other applications.

10.3.2.4.3 Pyrometallurgical Processes

The pyrometallurgical processing of metals includes the melting and refining of materials to produce pure metals, such as copper and lead. The refining steps produce secondary phases that are rich in other metals, like precious metals, which are defined in specific equipment (Oliveira 2012).

The decomposition of organic material using pyrometallurgical processes is an appropriate way to recover metals since most organic compounds decompose and become volatile at high temperatures. However, this process may release halogen compounds, dioxins, and furans, also producing scum due to the presence of ceramic materials and glass in waste. Besides, pyrometallurgical processes are not very efficient to recover tin (Sn), lead (Pb), and present other disadvantages (Blazsó et al. 2002; Veit 2005; Kasper 2011; Oliveira 2012).

10.3.2.4.4 Hydrometallurgy

The word “hydrometallurgy” denotes extraction and solubilization processes of metals in an aqueous medium, which are followed by the treatment of resulting solutions until valued forms of metals are obtained (Oliveira 2012). In these processes, attack solutions are used to dissolve and separate solid materials. These solutions are then submitted to processes like solvent extraction, precipitation, and refining to improve the selection of materials (Veit 2005; de Moraes 2011).

For Silvas (2014), several hydrometallurgical techniques may be used to treat rare earth elements, like chemical leaching with cyanide, thiosulfate, ligands (ethylenediaminetetraacetic acid (EDTA), diethylenetriaminepentaacetic acid (DTPA), nitrile acetic acid (NTA), oxalate, sodium hypochlorite, as well as sulfuric, hydrochloric, nitric acids, and aqua regia). Hydrometallurgical pickling with organic solvents, iron chloride, copper chloride, and hydrochloric acid is also used, besides bioleaching, which in turn consumes little energy, as microorganisms carry out extraction processes naturally. Bacterial leaching, also called biohydrometallurgy, is a process based on the solubilization of components in a mineral sample by bacteria (Brandl 2001; Garcia 1989; Garcia and Urenha 2001). This process uses bacteria to extract valuable elements from ores or waste.

A few chemical factors play an important role in these processes, like the pH of solutions, the composition of materials, solid-to-liquid ratio, and redox potential. Physical factors at play include porosity, particle size and shape, and permeability, affecting the speed at which constituents are transported from the solid to the liquid phase (Leaf 2017).

10.4 Conclusion

The amount of electrical and electronic equipment produced in the world is very high, and the market for these devices continues to expand, increasing the demand for elements extracted from natural environments such as metals and rare earth elements. However, several of these elements are geologically scarce, pointing to the need to adopt recovery methods (mechanical, pyrometallurgical, and hydrometallurgical processes) to obtain them from E-waste. In turn, this need means that urban mining may become an important factor in the sustainability of the production of electrical and electronic equipment.

References

ABDI – Agência Brasileira de Desenvolvimento Industrial (2013) Logística Reversa de Equipamentos Eletroeletrônicos – Análise de Viabilidade Técnica e Econômica. http://www.abdi.com.br/Estudo/Logistica%20reversa%20de%20residuos_pdf. Accessed 12 July 2017

ABINEE – Associação Brasileira da Indústria Elétrica e Eletrônica (2009) A Indústria Elétrica e Eletrônica em 2020. http://www.abinee.org.br/programas/imagens/2020a.pdf. Accessed 19 June 2018

Araújo MG (2013) Modelo de avaliação do ciclo de vida para a gestão de resíduos de equipamentos eletroeletrônicos no Brasil. Thesis, Universidade Federal do Rio de Janeiro, Rio de Janeiro, Brazil

Arndt NT, Fontboté L, Hedenquist JW, Kesler SE, Thompson JFH, Wood DG (2017) Future Global Mineral Resources. http://www.geochemicalperspectives.org/wp-content/uploads/2017/05/v6n11.pdf. Accessed 25 Jan 2018

Arora R, Paterok K, Banerjee A, Saluja MS (2017) Potential and relevance of urban mining in the context of sustainable cities. IIMB Manag Rev 29:210–224

Ayres RU, Peiró LT (2013) Material efficiency: rare and critical metals. Phil Trans R Soc A 371:20110563

Baccini P, Brunner P (2012) Metabolism of the anthroposphere: analysis evaluation design, 2nd edn. MIT Press, Cambridge, MA. ISBN 978-0-262-01665-0

Blazsó M, Czégény ZS, Csoma CS (2002) Pyrolysis and debromination of flame retarded polymers of electronic scrap studied by analytical pyrolysis. J Anal Appl Pyrolysis 64:249–261

Brandl H (2001) Microbial leaching of metals. In: Microbial diversity in bioleaching environments, chapter 8. Zurich, pp 192–217

BRASIL. Lei nº 12.305, de 2 de agosto de 2010. Institui a Política Nacional de Resíduos Sólidos (PNRS), altera a Lei no 9.605, de 12 de fevereiro de 1998; e dá outras providências. Disponível em:. Acesso em 19 mar. de 2019.

Britannica Academic Encyclopædia Britannica (2016) Electronic Waste. https://www.britannica.com/technology/electronic-waste. http://www.epa.ie/enforcement/weee/electrical and electronicequipment/. Accessed 25 Jan 2018

CETEM – MCT. Centro de Tecnologia Mineral. Ministério da Ciência e Tecnologia (2004) Tratamento de Minérios. 4a edição. Rio de Janeiro.

CETEM – Centro de Tecnologia Mineral (2015) Extração de ouro a partir de placas de circuito impresso por cianetação intensiva. Série Tecnologia Ambiental – ISSN 2015:0103–7374

Cobbing M (2008) Toxic tech: not in our backyard. Uncovering the Hidden Flows of e-waste. https://www.greenpeace.org/norway/Global/norway/p2/other/report/2008/toxic-tech-not-in-our-backyar.pdf. Accessed 19 June 2018

Cossu R, Williams ID (2015) Urban mining: concepts terminology challenges. Waste Manag 45:1–3

Cucchiella F, D'Adamo I, Koh SCL, Rosa P (2015) Recycling of WEEEs: an economic assessment of present and future e-waste streams. Renew Sust Energy Rev 51:263–272

Cui J, Forssberg E (2003) Mechanical recycling of waste electric and electronic equipment: a review. J Hazard Mater B99:243–263

da Silveira TA, Dorneles KO, Moraes CAM, Brehm FA (2016) Caracterização de placas de circuito impresso de smartphones orientada para reciclagem. Paper presented at the fifth innovation and technology seminar IFSul, 8–10 November 2016, ISSN 2446-7618

de Moraes VT (2011) Recuperação de metais a partir de processamento mecânico e hidrometalúrgico de placas de circuito impresso de celulares obsoletos. Thesis, USP, São Paulo, Brazil

Di Maria F, Micale C, Sordi A, Cirulli G, Marionni M (2013) Urban mining: quality and quantity of recyclable and recoverable material mechanically and physically extractable from residual waste. Waste Manag 33(12):2594–2599

Environmental Protection Agency (EPA) Electrical and electronic equipment. http://www.epa.ie/enforcement/weee/electricalandelectronicequipment/. Accessed 6 Feb 2018

EPA (United States Environmental Protection Agency) https://searchepagov/epasearch/epasearch?querytext=recycle+&typeofsearch=epa&doctype=all&originalquerytext=recycle+concept&areaname=&faq=true&site=epa_default&filter=&fld=&sessionid=9D9809E31912F3508726BA7629E6E5D2&prevtype=epa&result_template=2colftl&stylesheet=. Accessed 7 Feb 2018

European Commission (2010) Critical raw materials for EU Report of the Ad-hoc Working Group on Defining Critical Raw Materials. https://ec.europa.eu/growth/tools-databases/eip-raw-materials/en/system/files/ged/79%20report-b_en.pdf. Accessed 29 Jan 2018

European Commission 2014 Report on the critical raw materials for EU Report of the Ad-hoc Working Group on Defining Critical Raw Materials. https://ec.europa.eu/docsroom/documents/10010/attachments/1/translations/en/renditions/pdf. Accessed 29 Jan 2018

Garcia JRO (1989) Estudos da Biolixiviação de Minérios de Urânio por *Thiobacillus ferrooxidans*. Thesis, Instituto de Biologia Campinas, UNICAMP, São Paulo, Brazil

Garcia JRO, Urenha LC (2001) Lixiviação Bacteriana de Minérios. In: Lima UA, Aquarone E, Borzani W, Schmidel WS (eds) Biotecnologia Industrial-Processos Fermentativos Industriais, 1ed. Edgard Blücher Editora, São Paulo, pp 485–512

Garlapati VK (2016) E-waste in India and developed countries: management recycling business and biotechnological initiatives. Renew Sust Energy Rev 54:874–881

Gerbase A, Oliveira C (2012) Reciclagem do lixo de informática: Uma nova oportunidade para a Química. Quim Nova 35(7):1486–1492

Ghosh B, Ghosh MK, Parhi P, Mukherjee PS, Mishra BK (2015) Waste printed circuit boards recycling: an extensive assessment status. J Clean Prod 94:5–19

Gouveia AR (2014) Recuperação de metais de placas de circuito impresso por via hidrometalúrgica. Dissertation. Universidade do Porto, Lisboa, Portugal

GSMA – GSM Association (2015) eWaste in Latin America: Statistical analysis and policy recommendations. https://www.gsma.com/publicpolicy/ewaste-latin-america-statistical-analysis-policy-recommendations/gsma2015_report_ewasteinlatinamerica_english-2. Accessed 19 June 2018

Hayes PC (1993) Process principles in minerals and materials production. Hayes Publishing CO, Brisbane, p. 29

Henckens MLCM, Van Ierland EC, Driessen PPJ, Worrell E (2016) Mineral resources: geological scarcity, market price trends, and future generations. Resour Policy 49:102–111

Jaiswal A, Samuel C, Patel BS, Kumar M (2015) Go green with WEEE: eco-friendly approach for handling e-waste. Proc Comp Sci 46:1317–1324

Jo HJ, Kang HY, Lee IS, Jun YS (2017) Estimation of potential quantity and value of used and in-use stocks or urban mines in Korea. J Mater Cycles Waste Manag. https://doi.org/10.1007/s10163-017-0655-8. Accessed 12 July 2017

Kasper AC (2011) Caracterização e Reciclagem de Materiais Presentes em Sucatas de Telefones Celulares. Dissertation, Universidade Federal do Rio Grande do Sul, Porto Alegre, Brazil

Knoth R, Brandstotter M, Kopacek B, Kopacek P (2000) Automated disassembly of electr(on)ic equipment. Conference Record 2002 IEEE International Symposium on Electronics and the Environment

Krook J, Baas L (2013) Getting serious about mining the technosphere: a review of recent landfill mining and urban mining research. J Clean Prod 55:1–9

LEAF – Leaching Environmental Assessment Framework/Vanderbilt University Leaching Process. http://www.cresp.org/cresp-projects/waste-processing-special-nuclear-materials/leaf-leaching-environmental-assessment-framework/. Accessed 12 July 2017

Lederer J, Laner D, Fellner J, Recheberger H (2014) A framework for the evaluation of anthropogenic resources based on natural resource evaluation concepts. Paper presented at the SUM 2014, 2nd Symposium on Urban Mining, IWWG – International Waste Working Group, Bergamo, Italy. 16–21 May, 2014

Loureiro LFE (2013) O Brasil e a reglobalização da indústria das terras raras. CETEM/MCTI, Rio de Janeiro

Lundgreen K (2012) The global impact of e-waste: addressing the challenge. International Labour Office Programme on Safety and Health at Work and the Environment (SafeWork) Sectoral Activities Department (SECTOR). ILO, Geneva

Nicolai FNP (2016) Mineração urbana: avaliação da economicidade da recuperação de componentes ricos em Au a partir de resíduo eletrônico (e-waste). Thesis, Universidade Federal de Ouro Preto, Minas Gerais, Brazil

OLIVEIRA, P.C.F. Valorização de Placas de Circuito Impresso por Hidrometalurgia. Lisboa, 320 p., 2012. Tese (doutorado) – Universidade Técnica de Lisboa.

Palmiere R, Bonifazi G, Serranti S (2014) Recycling-oriented characterization of plastic frames and printed circuit boards from mobile phones by electronic and chemical imaging. Waste Manag 34:2120–2130

Parajuly K, Habib K, Liu G (2017) Waste electrical and electronic equipment (WEEE) in Denmark: flows, quantities and management. Resour Conserv Recycl 123:85–92

Rare Element Resources (2016) Rare Earth Elements. Littleton. Acessed 10 Mar 2019

Ribeiro PPM (2013) Concentração de metais contidos em placas de circuito impresso de computadores descartados. Projeto de graduação. Universidade Federal do Rio de Janeiro, Rio de Janeiro, 66 p

Richter D (2009) Uma rota de recuperação de metal a partir de escória secundária da produção de ferroníquel. Dissertação de Mestrado. USP, São Paulo,

Santanilla AJM (2012) Recuperação de Níquel a partir do Licor de Lixiviação de Placas de Circuito Impresso de Telefones Celulares. Dissertação (mestrado). São Paulo – USP

Sarath P, Bonda S, Mohanty S, Nayak SK (2015) Mobile phone waste management and recycling view and trends. Waste Manag 46:536–545

Sena FR (2012) Evolução da tecnologia móvel celular e o impacto nos resíduos eletroeletrônicos. Dissertation, PUC, Rio de Janeiro, Brazil

Shagun A, Kush A, Arora A (2013) Proposed solution of e-waste management. Int J Fut Comp Commun 2(5):490–493p

Silvas PCS (2014) Utilização de hidrometalurgia e biohidrometalurgia para reciclagem de placas de circuito impresso. Thesis, Universidade de São Paulo, São Paulo, Brazil

Song Q, Li J (2014) Environmental effects of heavy metals derived from the e-waste recycling activities in China: a systematic review. Waste Manag 34:2587–2594

Statista, The Statistics Portal (2017) Countries with the largest smelter production of aluminum from 2010 to 2017 (in 1000 metric tons). https://www.statista.com/statistics/264624/global-production-of-aluminum-by-country/. Accessed 13 Feb 2018

STEP – Solving the E-Waste Problem – Green Paper (2015) E-waste prevention take-back system design and policy approaches, ISSN: 2219-6579 (online)

Tansel B (2017) From electronic consumer products to e-waste: global outlook waste quantities recycling challenges. Environ Int 98:35–45

Tesfaye F, Lindberg D, Hamuyuni J, Taskine P, Hupa L (2017) Improving urban mining practices for optimal recovery of resources from e-waste. Min Eng 111:209–221

Tunsu C, Petranikova M, Gergoric M, Ekberg C, Retegan T (2015) Reclaiming rare earth elements from end-of-life products: a review of the perspectives for urban mining using hydrometallurgical unit operations. Hydrometallurgy 156:239–258

Umicore (2016) UMICORE BRASIL LTDA. Processamento de sucata eletrônica. V Seminário Internacional Sobre Resíduos de Equipamentos Eletroeletrônicos – 18 e 19 de agosto de 2016, Recife, Pernambuco, Brasil

UNEP – United Nations Environment (2015) Waste crime – waste risks: gaps in meeting the global waste challenge http://web.unep.org/ourplanet/september-2015/unep-publications/waste-crime-waste-risks-gaps-meeting-global-waste-challenge-rapid. Accessed 6 Feb 2018

UNEP. – United Nations Environment Programme (2009) Recycling – from e-waste to resources. http://www.unep.fr/shared/publications/pdf/DTIx1192xPA-Recycling%20from%20ewaste%20to%20Resources.pdf. Accessed 19 June 2018

União Europeia (2012) Directive 2012/19/UE of the European Parliament and of the Council of 4 July 2012 on waste electrical and electronic equipment (WEEE). Off J L 197:38–71. https://eur-lex.europa.eu/LexUriServ/LexUriServ.do?uri=OJ:L:2012:197:0038:0071:en:PDF. Accessed 5 Feb 2018

USGS – (United States Geological Survey) (2016) Iron and steel production. https://minerals.usgs.gov/minerals/pubs/commodity/iron_&_steel/mcs-2016-feste.pdf. Accessed 13 Feb 2018

Veit HM (2001) Emprego de Processamento Mecânico na Reciclagem de Sucatas de Placas de Circuito Impresso. Dissertation, UFRGS, Porto Alegre, Brazil

Veit HM (2005) Reciclagem de Cobre de Sucatas de Placas de Circuito Impresso. Thesis, UFRGS, Porto Alegre, Brazil

Ventura EACC (2014) Estudos de processos físicos para recuperação de metais de placas de circuito impresso. Dissertation, Universidade do Porto, Porto, Portugal

Vivas R de C, Costa FP (2013) Tomada de decisão na escolha do processo de reciclagem e recuperação de metais das placas eletrônicas através de análise hierárquica. IV Congresso Brasileiro de Gestão Ambiental. Salvador/BA.

Wen Z, Zhang C, Ji X, Xue Y (2015) Urban Mining's potential to relieve China's coming resource crisis. J Ind Ecol 10(6):1091–1102

World Steel Association (1978) Handbook of world steel statistics 1978. https://webarchiveorg/web/20150513202042/ https://www.worldsteel.org/en/dam/jcr:09ed87af-366e-44c6-9f55-63c09f76b520/A%2520handbook%2520of%2520world%2520steel%2520statistics%25201978.pdf. Accessed 13 Feb 2018

World Steel Association (2010) Steel statistical yearbook. https://www.worldsteel.org/en/dam/jcr:1ef195b3-1a46-41c2-b88b-6072c2687850/Steel+statistical+yearbook+2010.pdf. Accessed 13 Feb 2018

World Steel Association (2011) Steel statistical yearbook. https://www.worldsteel.org/en/dam/jcr:c12843e8-49c3-40f1-92f1-9665dc3f7a35/Steel%2520statistical%2520yearbook%25202011.pdf. Accessed 13 Feb 2018

World Steel Association (2012) Steel statistical yearbook. https://www.worldsteel.org/en/dam/jcr:a0d5110b-80e1-4f1d-a6f0-a9054c07b672/Steel%2520Statistical%2520Yearbook%25202012.pdf. Accessed 13 Feb 2018

World Steel Association (2013) Steel statistical yearbook. https://www.worldsteel.org/en/dam/jcr:7bb9ac20-009d-4c42-96b6-87e2904a721c/Steel Statistical-Yearbook-2013.pdf. Accessed 13 Feb 2018

World Steel Association (2014) Steel statistical yearbook. https://www.worldsteel.org/en/dam/jcr:c8b1c111-ce9b-4687-853a-647ddbf8d2ec/Steel-Statistical-Yearbook-2014.pdf. Accessed 13 Feb 2018

Yu J, Welford R, Hills P (2006) Industry responses to EU WEEE and ROHS directives: perspectives from China. Corp Soc Respons Environ Manag 13:286–299

Chapter 11
Sustainable Electronic-Waste Management: Implications on Environmental and Human Health

K. Grace Pavithra, Panneer Selvam Sundar Rajan, D. Balaji, and K. P. Gopinath

Contents

Abstract The increasing level of Electronic-waste and its improper disposal and unsafe treatment pose significant risks to the environment and human health. They raise several challenges to the sustainable development goals. Electronic-waste is considered one of the fastest-growing pollution problems all over the world as per the United Nations environment programme estimates. This rapid growth is influenced by planned product extinction, lower prices, and change of lifestyle. Unfortunately, a major amount of Electronic-waste is recycled in the informal sector

K. Grace Pavithra · P. S. Sundar Rajan · D. Balaji · K. P. Gopinath (✉)
Department of Chemical Engineering, SSN College of Engineering, Chennai, Tamil Nadu, India
e-mail: gopinathkp@ssn.edu.in

A. Khan et al. (eds.), *E-waste Recycling and Management*, Environmental Chemistry for a Sustainable World 33, https://doi.org/10.1007/978-3-030-14184-4_11

and results in toxic exposures to the recyclers, especially to women and children. Electronic-waste consists of valuable metals as well as environmental contaminants especially polybrominated diphenyl ether and polychlorinated biphenyls. The chemical composition of Electronic-waste changes with the innovation of new technologies and pressure from environmental organizations. As the reprocessing and recycling technologies with minimal environmental impacts are found to be expensive, rich countries export unknown quantities of Electronic-waste to developing countries, where recycling techniques including burning and dissolution in strong acids result in localized contaminations of water and food chains. This chapter deals with the generation of electronic-waste and its disposal pathways, and it especially covers the various contaminants that affect human health as well as our environment.

Keywords Industrialization · Electronic-waste · Sustainable development goals · Human health · Environment

11.1 Introduction

Due to rapid industrialization, industries have moved towards an environment with the reduction of human power and an increase in usage of electrical as well as electronic equipment. The innovation of electrical and electronic products leads the way to its usage in manufacturing and in another industrial sector. The development of faster and reliable processing technologies such as mobile phones, computers, and laptops drives consumer to use current and upgraded technology products with no longer use of older products. This development, in turn, gives a tremendous increase in the usage of electronic-waste products among consumers. Step Initiative (2014) framed a definition for Electronic-waste as follows: "Electronic waste is a term used to cover items of all types of electrical and electronic equipment and its part that have been discarded by the owner as waste without the intention of re-use."

It is considered as a fast-growing waste sector, which includes a large variety of components that include toxic as well as resourceful products. In the field of information and communication technology due to rapid advancement, substitution, and miniaturization, Electronic-waste increases day by day. Electronic-waste is also referred to as waste electrical and electronic equipment.

Baldé et al. (2015) categorized electrical and electronic equipment into six types; they are as follows:

- Temperature exchange equipment – referred to as cooling and freezing equipment. The equipment such as refrigerators, air conditioner, heat pump, and freezers will come under this category.
- Screens and monitors – monitors, laptops, tablets, and television.
- Lamps – light-emitting diode lamps, fluorescent lamps, and discharge lamps.
- Large equipment – washing machine, cloth dryers, electric stoves, photovoltaic panels, printing, and copying equipment.

- Small equipment – vacuum cleaners, microwaves, ventilation equipment, electric kettles, radio, video recorder, calculators, electronic toys, medical devices, and electronic tools.
- Small information technology and telecommunication equipment – mobile phones, global positioning system, personal computers, printers, and telephones.

The lifetime of the six different categories varies according to their quantities, economic values, and impact on environmental health if recycled improperly. The consumer mentality in disposing of the electrical and electronic equipment, as well as the recycling technologies, also varies accordingly. It is noted that in 2014, 41.8 million metric tons of Electronic-waste was produced as waste, and in near 2018 it will increase to 50 metric tons (Baldé et al. 2015). From 1992 to 2005, the lifespan of computers shrunk from 4.5 years to 2 years. This is due to the upgradation of newer and latest version with the old ones by the consumers. Unsuitable recycling makes the electronic-waste as potential environmental and health hazards. It is noted that Electronic-waste consists of 60 metals, which include copper, silver, palladium, platinum, gold, etc. These metals when recovered could reduce the demand for new metal production to some extent. The production of hazardous by-products in electrical and electronic equipment is called an informal sector of Electronic-waste. Inadequate recycling technologies and manpower were seen in terms of Electronic-waste. Unexpected exposure to Electronic-waste especially to children is found to be vulnerable. In this chapter we will be focusing on the electrical and electronic equipment, their impact to our environment as well as to health, recycling of electrical and electronic equipment, approaches in the reduction of health effects, and special attention towards the areas of improvement.

11.1.1 Electronic-Waste and Sustainable Developmental Goals

In order to end poverty and to give property assurance and protection of our planet for the next 15 years, in September 2015, the United Nations and state members adopted 2030 agenda for sustainable development. It consists of 17 sustainable development goals and 169 targets. From the goals, it is concluded that Goal 3: Good health and well-being, focuses on death rates and illness caused by hazardous chemicals and contamination of air, water, and soil. Goal 6: Clean water and sanitation and, seeks to achieve safe drinking water and to reduce pollution and dumping of hazardous waste. Goal 8: Decent work and economic growth aims at entrepreneurship, innovation, and micro-, small-, and medium-sized enterprises. Goal 11: Sustainable cities and communities focuses on per-capita environmental impact. Goal 12: Responsible consumption and production aimed at the life cycle of individual products and their release to the soil, water, and air, and on waste generation reduction by prevention, reduction, repair, recycling, and reuse. Goal 14: Life below water goes well with the management of Electronic-waste. Electronic-waste when treated improperly creates health issues and contaminates air, water, and soil, and untrained people in dismantling process provides risk to the people as well as to the

Fig. 11.1 Sustainable development goals related to environment and health

environment (Baldé et al. 2017). Representation of sustainable developmental goals highlighting the goals related to human and environmental health is mentioned in Fig. 11.1.

11.2 Electronic-Waste Tracking and Driving Trends

From 2000 to 2016, there is a rapid growth in consumption of electrical and electronic equipment. The products such as fridges, washing machines, electric furnace, and television found tremendous growth over this period. At present, the countries, which are in the economic cooperation and development organization, have a high demand for electrical and electronic equipment. Three methods are predicted (Lohse et al. 1998) for the estimation of Electronic-waste, and they are:

- The consumption and use method; it averages the household electrical and electronic equipment for the prediction of quantities of waste electrical and electronic equipment.
- The market supply method, which uses data about sales and production in the specified region.

For the first two methods, the weight and the lifetime of the electrical and electronic equipment have to be noted, and the third method is Swiss environmental agency estimates, which determines the saturated position in household, i.e. for one new appliance, an old one is replaced. This method is not applicable practically as it is not the case in all households.

The rapid expansion of mobile and broadband networks allows people to access the Internet in rural areas and areas which are unconnected previously. The heavy competition in the telecommunication market for mobile brands is a key factor for a decrease in price in electrical and electronic equipment, and this makes the baseline for the spread of electrical and electronic equipment and the Internet drastically. A single person owns more than one mobile phone, and many people use various other electronic gadgets to access internet. Due to the higher speed and upgradation to the latest technologies, the consumer changes their laptops, personal computers, television, and other devices regularly. The older equipment is considered outdated, and people are replacing it when it is not damaged or broken. For example, from analogue to digital conversion, many televisions were replaced with digital signals, which lead the way to carbon-ray-tube television invasion (ITU 2015, 2017). The UN conference on trade and development estimated in 2015:

- The e-commerce global business-to-business goes beyond US$22 trillion, and business-to-consumer exceeded US$3 trillion.
- It is concluded that 40% of enterprises receive orders using the Internet.

11.2.1 Electronic-Waste Statistics and Measurement

Over the past five decades, there is an increase in the quantity of Electronic-waste generation in the United States, the European Union, and developing countries like India and China. As per the US Environmental Protection Agency, each household uses on an average 34 electronic devices and appliances which results in the generation of 5×10^6 tons of Electronic-waste per year. In the European Union, it was estimated that each citizen generates 15 kg of Electronic-waste per year, which approximately totals to 7×10^6 tons. Electrical and Electronic-waste contributes 8% of total municipal solid waste. In developing countries, the Electronic-waste generation per capita is around 1 kg per year, and it is found to be increasing rapidly. Import of Electronic-waste from developed countries also contributes a major amount of Electronic-waste to developing countries. In the process of recycling and reuse, 50–80% of waste electrical and electronic equipment generated in developed countries are shipped to developing countries, which are against international laws (Bastian et al). When Electronic-waste is not disposed of properly, it affects our environment and living beings. In developing countries, Electronic-waste is dumped into the soil without treatment due to the lack of strict regulations and financial resources leading to the soil as well as groundwater contaminations. In India, recycling happens in the form of informal sector where people tear apart the components in the electronic equipment without any safety measures. People working in this sector are exposed to toxins 24 hours a day as they live, sleep, and cook in the same place. In Delhi, illegal and dangerous recycling practices are adopted (Tsydenova and Bengtsson 2011, Needhidasan et al. 2014).

In Baldé et al. (2015) research, Electronic-waste progress has been discussed in an accurate manner. Some of the indicators are discussed in the above publication, and they are:

- Electrical and electronic equipment market scenario
- Electronic-waste totally generated
- Recycle of Electronic-waste done officially
- Collected Electronic-waste rate

The life cycle of Electronic-waste that is from electrical and electronic equipment and the Electronic-waste management are explained in four phases, and they are as follows:

- **Phase 1**
  Using apparent consumption method or using the statistics, which are maintained from the sales to the Electronic-waste registry, market entry phase can be analysed.
- **Phase 2**
  The sold product that enters the household or to the business comes under the stock phase. Using national-level surveys in the household as well as in business, the stock phase can be analysed. The stock phase includes sales information on electrical and electronic equipment and the product residence time.
- **Phase 3**
  When the product becomes out of date to the owner, it ends up as waste. Generation of Electronic-waste becomes an important factor in Electronic-waste statistics.
- **Phase 4**
  Electronic-waste is managed using one of the following techniques, and they are:
  Official take-back system – In order to reduce the environmental impact under the national Electronic-waste legislation, designated organizations are allotted for the collection of Electronic-waste.
  Mixed residual waste collection – Consumers are directly involved in the disposal of Electronic-waste, and as an impact, the disposed Electronic-waste is made to treat with residual waste. Landfills and incineration are found to be

currently using the technique for residual waste, but for the disposed Electronic-waste, suitable techniques need to be imposed as landfill and incineration are found to release toxin leachate and air pollution that creates negative impact to our environment.

Collection outside the take-back system – In this scenario countries are divided based on the waste management procedure they follow which is developed and undeveloped. In a developed waste management system, Electronic-waste which is collected by the individual dealers is accounted, and the Electronic-waste end up locations such as metal recycling; plastic recycling is explored, and the information is not reported to the official take-back system to avoid double counting. In undeveloped countries, the process such as local dumping, export, and recovery of value-added substance is done. Fig. 11.2 Illustrates the life cycle of electrical and electronic equipment into Electronic-waste and the common Electronic-waste management scenarios

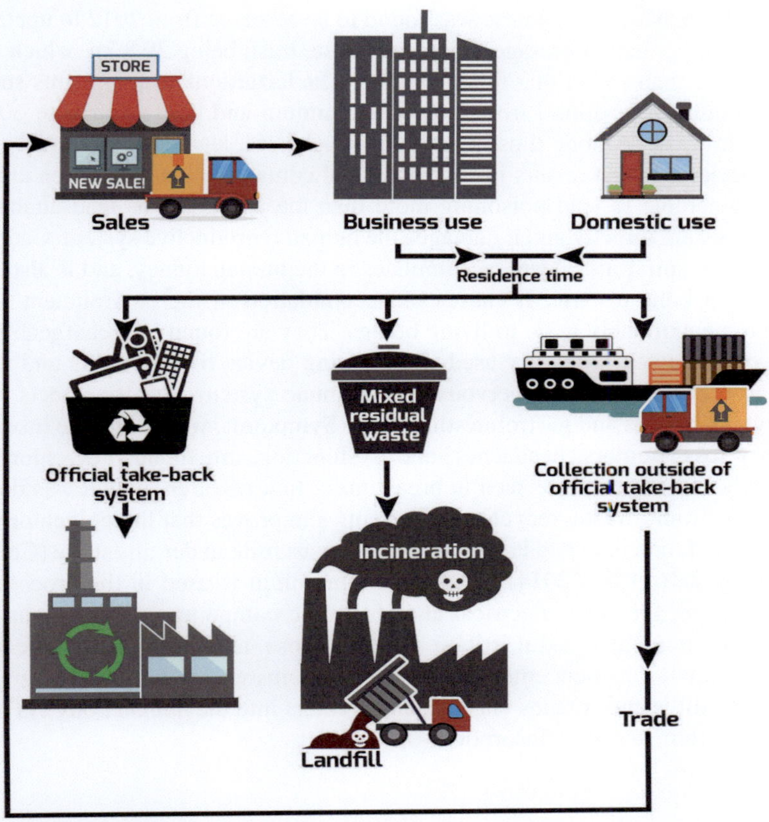

Fig. 11.2 The life cycle of electrical and electronic equipment into Electronic-waste and the common electronic-waste management scenarios

11.3 Positive and Negative Effects of Electronic-Waste

Positive Impact A research stated that (IBT 2012) 40–50 times of metals which are found to be precious are recovered from the Electronic-waste than from their native ores, and it was stated that for each year $21 billion of valuable metals are accumulated in Electronic-waste. The metals such as indium, palladium, silver, copper, and gold are often found in Electronic-waste. Forty per cent of valuable metals are recovered from printed boards (Golev et al. 2016), and it is stated that 320 tons of gold and 7500 tons of silver are used by electronic industries every year (BullionStreet 2012). In the Electronic-waste stream, notebooks, tablets, and smartphones are considered as valuable products due to the presence of high amount of precious metals (Cucchiella et al. 2015).

Negative Impact According to StEP in 2012 analysis, China and the United States contribute larger amount of Electronic-waste and waste electrical and electronic equipment with 11.1 million tons in China and 10 million tons from the United States. Now the scenario was found to be reversed from 2012 to upcoming years due to per capita average Electronic-waste trash being 29.8 kg, which is six times higher than China that figures 5.4 kg. The hazardous contaminants such as lead, cadmium, beryllium, iron, copper, aluminium and gold contribute 60% of Electronic-waste. Among those hazardous substances, lead is used in many electronic devices, which results in environmental contaminations. Children are normally vulnerable to lead poisoning more than the adults are as lead attacks the nervous system directly and it damages the human reproductive system. Cadmium affects the respiratory system, accumulates in the human kidney, and is also associated with bone disease. It causes bioaccumulation in the environment and is found to be extremely toxic to living beings. They are found in rechargeable batteries of computers. Mercury used as a lighting device like monitors and televisions affects the digestive, nervous, and immune systems. It also affects lungs, kidneys, skin, eyes and gastrointestinal tract. Symptoms, which include insomnia, memory loss, tremors, headaches, and dysfunction, are found. Polychlorinated biphenyls and mercury are seen in breast milk. In a research, which was done in China Electronic-waste, recycling processing site proves that the polychlorinated biphenyl and mercury level is seen higher in breast milk under this study (Ceballos and Dong 2016, UNU 2014). Hexavalent chromium is used in the process of a metal housing for many electrical and electronic components. Hexavalent chromium is a carcinogen, and it irritates the lung, nose, and throat. Electronic-waste with or without treatment enters into aquatic systems via leaching from dumpsites or from landfills and spreads via dust, and it enters into the human body via ingestion, inhalation, and skin absorption.

11.4 Products that Make Challenges to a Recycler

11.4.1 Solar Panel

In private as well as in industrial markets, solar panels are considered as generally known product worldwide. The polyvinyl panels installed in 1990's are on the verge of expiry which may not find a suitable recycling factility. Polyvinyl panels are based on silicon and a small number of rare earth metals such as cadmium, tellurium, indium, or gallium, which mostly ends in landfills, and 10% of polyvinyl panels are recycled in recycler's point of view. In this case, polyvinyl panels are not considered as environmental contaminants. Some researchers demonstrated that in the next few years, there will be a vast dumping of polyvinyl panels due to the installation of the massive amount of panels currently (Choi and Fthenakis 2014, Bustamante and Gaustad 2014).

11.4.2 Liquid Crystal Display and Cathode-Ray Tube Monitors

According to European Union waste electrical and electronics, cathode-ray tube and liquid crystal display technologies are considered hazardous substance. In liquid crystal display, panels and the backlights, which contain mercury, are considered as hazardous. The UK Environmental Agency in 2010 stated that "liquid crystal displays", which do not contain mercury backlights, are considered as non-hazardous. From the statement, it is concluded that key elements which make liquid crystal display a hazardous component are backlight containing mercury, and other hazardous substances are polyvinyl butyral and brominated flame retardants. It consists of three types of glass and the electron gun. As the image formation process occurs within a cathode-ray tube, it is considered as the main component in the monitor. In addition, cathode-ray tube is considered as the heaviest component comprising of 60% of the total weight in the monitor (Veit and Bernardes 2015), and it is composed of 85% of glass, which is located in the front panel (65%), funnel (30%), and neck glass (5%). Silica, alumina, lime, magnesia, boric acid, etc. (Herat 2008) are considered as main components of glass. In order to capture the escape of radiation from the tube, lead is added to the glass, and the panel glass contains barium. The lead content in the glass also makes glass as a hazardous material. The glasses in the cathode-ray tube are processed to bricks, nuclear waste encapsulation, construction aggregates, decorative tile, fluxing agent, and sandblasting medium (Guo et al. 2010a, b). The leaching of lead, barium, and other toxic materials present in the glass causes environmental risk. Leads are found to be toxic to kidneys, and its accumulation in our body affects the nervous and reproductive systems. High dosage results in haemorrhage and brain oedema. Nearly 40% of lead is from landfills, which are from electrical and electronic equipment (UNEP 2014).

11.4.3 Printed Circuit Boards

They are employed in computers as well as in electronic equipment for mechanical support, i.e. to connect the electronic components electrically by means of conductive pathways. Printed circuit systems place an essential role in all electric and electronic equipment. By means of additive and subtractive methods, circuit patterns are created. It works as a platform for mounting semiconductor chips and capacitors. The printed circuit boards contribute 3% of Electronic-waste in the overall Electronic-wastes generated. The materials such as aluminium, copper, iron and steel, lead, zinc, paper, and plastics are normally found in printed circuit boards. Metal, ceramic, and plastic compositions are found to be 40%, 30%, and 30%, respectively. Some concentrations of precious metals such as gold and palladium are found richer than their natural ores. Recycling process such as mechanical, automated, and semiautomated approaches is followed. The mechanical approach involves two steps; the first step consists of the separation of different components and materials, and the second step involves further separation and processing (Hall and Williams 2007, Guo et al. 2010a, b). In automated approach the use of image processing identifies the reusable parts in printed circuit boards by comparing the shape and labels in the database which are gathered from the manufacturers and from the reuse market. The reusable products and the hazardous components are separated in dismantling cell. The semiautomated approach is found to be flexible. The electronic components are removed by a combination of heating above the solder melting point. Semiautomated dismantling cell was connected for separating the reusable and hazardous component (Yi et al. 2007).

11.4.4 Cooling and Freezing Appliances

The release of chlorofluorocarbon is seen due to improper disposal of cooling and freezing appliances. It is reported that more than three-fractional amount of ozone-depleting substance is released from refrigerants and foams in cooling and freezing appliances. Scarp steel, compressors, cables, plastics, glass, and oil are the material compositions of refrigerators. Ferrous and non-ferrous metals account for 50% and 8% and plastics account for 20–25%. Polystyrene has high value in the secondary market and used as an insulating material in refrigerators and freezers. New cooling and blowing agents with less impact on the surroundings have to be introduced to the field of refrigerators and freezers.

11.4.5 Batteries

Batteries consist of one or more than one electrochemical cells, which are connected, in series as well as in parallel in order to produce electrical energy. The cell consists of anode, cathode, and electrolyte. Batteries play a major role in electrical

Fig. 11.3 Many people own multiple devices

and electronic devices in mobile phones, laptops, toys, cordless phones, personal computers, etc. The batteries are to be disassembled and recycled in a specific manner. Spent batteries are the secondary source of metals with high concentration levels. Metals such as cobalt and nickel are found in lesser amounts in nickel-cadmium, nickel-metal hydride, and lithium-ion batteries. Safe disposal includes landfills, stabilization, incineration and other recycling processes. Incineration of batteries is found to be expensive, and some toxins like mercury, cadmium, and dioxin are emitted during the incineration. Two main steps are involved in recycling, and they are waste preparation and metallurgical processing. Figure. 11.3 illustrates the use of multiple electronic devices by many people.

11.5 Electronic-Waste to Human Health and Environment

Electronic-waste may consist of 60 different elements, which are hazardous as well as non-hazardous in nature. Electrical and electronic equipment are considered major consumers of precious metals, and huge global demand is created due to the consumption. A simple mobile phone may consist of 40 elements from a periodic table including metals like copper (Cu), tin (Sn), cobalt (Co), indium (In), and antimony (Sb) and precious metals, which include silver (Ag), gold (Au), and palladium (Pd). One ton of phone handsets contains 3.5 kg of Ag, 340 g of Au, 140 g of palladium, and 130 kg of copper, and the remaining are filled with plastic and ceramic materials. The lithium-ion battery in the phone consists of approximately 3.5 kg of cobalt. In addition to electrical household appliances, batteries, capacitors, cathode-ray tubes, glass, etc. are considered Electronic-waste. In several countries recycling of above-mentioned Electronic-waste is done formally as well as in an informal manner. Formal recycling techniques utilize well-designed techniques and machinery to separate the useful products from the Electronic-waste in an expansive manner, and this type of formal recycling techniques does not have much impact to our environment when compared to informal recycling techniques, which are carried in underdeveloped and developing countries, which emit several pollutants in the environment. The exposure to hazardous components in the way of ingestion, skin contact, and inhalation by the mediums like soil, water, food, and air is seen. The risk of exposure is seen in workers; women and children are found to be at maximum risk due to the additional medium of exposures (maternal feeding, hand-to-mouth activities, care fewer behaviours). The exposure from the workers transfers to other family members through dermal contact, clothes, etc.; hazardous chemicals or compounds from Electronic-waste come either from the recycling process or from components of the electronic equipment. Some persistent organic pollutants such as brominated flame retardants, polychlorinated biphenyls, hexabromocyclododecane, polybrominated diphenyls, di-brominated diphenyl ethers, and polychlorinated and polybrominated dioxins are seen in Electronic-waste. Persistent organic pollutants formed during dismantling and smelting consist of polychlorinated dibenzofurans, polychlorinated biphenyls, and polychlorinated dibenzodioxins. Polycyclic aromatic hydrocarbons, which are generated during incomplete combustion of fuels like coal, gas, oil, etc., react with Electronic-waste materials and form hydrocarbons. Heavy metals such as lead, cadmium, chromium, mercury, copper, manganese, nickel, arsenic, zinc, iron, and aluminium are the results of Electronic-waste threats. Table 11.1 briefly shows various hazards caused by different components of Electronic-waste.

11.5.1 Effects on Air

Many Electronic-waste contaminants spread into the air via dust by transporting the Electronic-waste to developing countries where recycling processes are poorly regulated. In developing countries, due to informal economics, Electronic-wastes are recycled by way of dismantling and shredding with the release of a large amount of dust, which creates a respiratory problem for workers. Unregulated burning at low temperatures releases dioxins, which damages humans as well as animals.

11.5.2 Effects on Aquatic Organisms

Introduction of Electronic-waste contaminants to water bodies poses hazards to aquatic organisms. The local resident's life was affected due to dumping and other Electronic-waste recycling activities. The heavy metals such as lead, barium,

Table 11.1 Electronic-waste exposures and its potential hazards

Pollutants	Components of the electrical and electronic equipment	Effects to humans		Source of exposure	Route of exposure	References
		Temporary	Permanent			
Polybrominated diphenyl ethers	Flame retardants, electronic components	Fatigue, headache, reduced capacity of work, dizziness, irritability in combination with gastrointestinal syndromes such as diminished appetite, weight loss, abdominal pain	Neurodevelopmental toxicity, thyroid hormone imbalance, liver tumours	Air, soil, sediments, humans, wildlife, sewage treatment plant biosolids	Ingestion, inhalation, transplacental	Siddiqi et al. (2003)
Polybrominated biphenyls						
Polychlorinated biphenyls						
Polycyclic aromatic hydrocarbons	Incomplete combustion	Lung cancer		Atmosphere, surface soils	Ingestion, inhalation, and transplacental	Hussein et al. (2016)
Polychlorinated dibenzofurans, pyrene, chrysene, fluoranthene, fluorene, phenanthrene, acenaphthene, anthracene						
Heavy metals						
Arsenic	Light-emitting diodes, doping material for Si, semiconductors, microwaves, solar cells	Cancers in lungs, bladder, and skin	Carcinogenic, chronic disease	Air, soil, and water		Kumar et al. (2017) and Grant et al. (2013)
Barium	Getters in cathode-ray tube screens, plastic and rubber fillers, electron tubes	Brain, heart, liver, and lung damage		Air, water, soil, and food	Ingestion, dermal contact, and inhalation	
Cadmium	Toners, batteries, plastics, circuit boards, monitor	Birth defect and heart, kidney, and lung damage	Nephrotoxicity	Air, dust, soil, water, and food	Ingestion and inhalation	

Pollutants	Components of the electrical and electronic equipment	Effects to humans		Source of exposure	Route of exposure	References
		Temporary	Permanent			
Chromium	Data tapes, floppy disk, switches, solar	Bronchitis	Carcinogenic	Air, dust, water, and soil	Inhalation and ingestion	
Cobalt	Insulator	Thyroid damage	Vision problems, heart problems, vomiting and nausea			
Copper	Conductor cables, coils, circuitry, wiring	Liver and kidney damage, Wilson's disease	Irritation in nose, mouth, headaches, dizziness, vomiting, diarrhoea			
Lead	Cathode-ray tube screens, transistors, lasers, light-emitting diode, thermoelectrical elements,	Toxic to human, plants, and animals	Fatigue, sleeplessness, arthritis, hallucinations, vertigo, headache, hypertension			
Lithium	Lithium batteries			Air, soil, water, and food	Inhalation, ingestion, and thermal contact	
Mercury	Fluorescent lamps, alkaline batteries	Damages nervous system	Bioaccumulation, neurological and behavioural changes	Air, vapour, water, soil, and food	Inhalation, ingestion	
Selenium	Older photocopying machines as photosensitive drums	Lung tissue damage, oxidative damage in tissues	Headaches, diarrhoea			
Beryllium	Motherboard	Lung cancer, berylliosis	Headache, diarrhoea	Dust, air, food, water	Inhalation, ingestion	
Rare earth elements	Fluorescent layer	Brain, heart, liver damage	Birth defects			

mercury, lithium etc. from Electronic-waste when improperly disposed leach to the soil and reach groundwater channels. Some of the heavy metals are found to be carcinogenic, and intake of contaminated water by humans and land animals creates major health issues.

11.5.3 Effects on Soil

Toxic heavy metals from Electronic-waste enter to humans by "soil-crop-food pathway", the effects of which are birth defects and brain, heart, liver, kidney, and skeletal system damage. When computer monitors and electronics are burned, cancer-producing dioxins are released into the air, and if thrown in landfills, the toxins may leach into groundwater. Table 11.2 represents the toxic substances and their health impacts.

Table 11.2 Common toxic substances associated with electronic-waste and their health impacts

S. No	Substances	Way of exposure	Health issues
1.	Antimony	Ingestion, inhalation, skin contact	Damages the blood, kidneys, lungs, nervous system, liver, and mucous membranes
2.	Cobalt	Inhalation, ingestion	Organ damage, toxic to lungs, and carcinogenic effects
3.	Gallium	Ingestion, inhalation, skin contact	Toxic to lungs, mucous membrane, severe exposure leads to death
4.	Copper	Inhalation, ingestion	Results in death, toxic to lungs and mucous membranes, organ damage
5.	Arsenic		Skin lesions, peripheral neuropathy, gastrointestinal symptoms, diabetes, renal system effects, cancer, and cardiovascular disease
6.	Cadmium	Inhalation, ingestion	Accumulation in kidney and liver, human carcinogen, and has toxic effects on the kidney, skeletal system, and respiratory system
7.	Dioxins		Toxic and cause chloracne, damage immune system, interfere with hormones and cause cancer
8.	Barium	Inhalation, ingestion	Short-term muscle weakness and damages heart
9.	Beryllium	Ingestion, skin contact	Carcinogenic, chronic beryllium disease, warts, etc.
10.	Chlorofluorocarbons	Contact, inhalation	Potent greenhouse gas. Direct exposure causes unconsciousness, irregular heartbeat, drowsiness, coughing, difficulty in breathing, sore throat, eye redness, and pain

Sources: Compendium (2016)

11.6 Electronic-Waste Management

In minimizing the negative effects on the environment and human health as well as to increase the recycling of Electronic-waste, Electronic-waste legislation is followed in 90 jurisdictions. European Union, Japan, and South Korea have enlightened legislation and controls, which are implemented, and some developing countries like China, India, and Brazil are taking initiating steps in implementing Electronic-waste legislation. In many countries, no regulations are followed since (Li et al. 2015). Since 2002 European Union has established some of the legislation directive on waste electrical and electronic equipment and restriction of hazardous substances such as directive on registration, evaluation, authorization, and restriction of chemical substances, and Japan also established some of the legislation like waste management and public cleansing law, home appliance recycling law, and small appliance recycling. In China, they have enacted environmental protection law of the People's Republic of China, Law of the People's Republic of China on Circular Economy Promotion, Law of the People's Republic of China on cleaner production promotion, etc. Moreover, in terms of Electronic-waste, they have enacted management regulation on the recycling of waste electrical and electronic products, opinions on strengthening the prevention and control of pollution from electronic-waste, and administrative measures for the prevention and control of environmental pollution from Electronic-waste (Zhou and Xu 2012). Developed countries are found to be different from developing countries in terms of Electronic-waste usage and disposal. The issues in developing countries are as follows (Heeks et al. 2015, Osibanjo and Nnorom 2007):

- Inadequate handling threats
- Lack of formal recycling systems
- Absence of or improper recycling legislation
- Rapid development and their Electronic-waste quantities in the information and communication sector
- The tremendous increasing rate of Electronic-waste generation
- Management issues found in component handling in Electronic-waste
- The negative impact of present Electronic-waste management strategies

11.7 Conclusion

Electronic-waste is found to be a serious issue at both local and global scales. Electronic-waste-related problems initially started in developed countries, and now it has extended to developing countries across the world. Rapid change in consumer technologies and innovations moves current technologies to extinction. If the end-of-life management of electrical and electronic equipment is not managed properly, it will result to release of toxic substances, which can contaminate our environment and threaten human health. Many studies from Electronic-waste recycling plant evidence of the release of toxic heavy metals as well as contaminants to our environment, and a major source of contaminants is seen in informal recycling sectors, and informal Electronic-waste recycling has long been accepted to lead to dangerous environmental pollution. The international health community, policy experts, and non-governmental organizations in a joint venture with the national government should create awareness among people, through the creation of policy solutions, conducting educational programmes, and setting up goals for the reduction of Electronic-waste exposure and its health effects.

References

Baldé CP, Wang F, Kuehr R, Huisman J (2015) The global electronic waste monitor — 2014

Baldé CP, Forti V, Gray V, Kuehr R, Stegmann O (2017) The global Electronic waste monitor-2017, Quantities, flows and resources, United Nations University (UNU), International Telecommunication Union (ITU) & International Solid Waste Association (ISWA), Bonn/Geneva/Vienna

BullionStreet (2012) Electronics industry uses 320 tons of gold, 7500 tons of silver annually., Retrieved May 6, 2016, from http://www.bullionstreet.com/news/electronics-industry-uses-320-tons-of-gold7500-tons-of-silver-annually/2255

Bustamante ML, Gaustad G (2014) The evolving copper–tellurium byproduct system: review of changing production techniques & their implications, pp 11–16

Ceballos DM, Dong Z (2016) The formal electronic recycling industry: challenges and opportunities in occupational and environmental health research. Environ. Internat. 95:157–166

Choi JK, Fthenakis V (2014) Crystalline silicon photovoltaic recycling planning: macro and micro perspectives. J Clean Product 66:443–449

Compendium of technologies for the recovery of materials from WEEE/Electronic waste, UN environment, 17 Oct 2016, http://web.unep.org/ietc/sites/unep.org.ietc/files/Electronicwaste%20Compendium%2017%20Oct%202016.pdf

Cucchiella F, D'Adamo I, Lenny Koh SC, Rosa P (2015) Recycling of WEEEs: an economic assessment of present and future electronic waste streams. Renew Sust Energ Rev 51:263–272

Golev A, Schmeda-Lopez DR, Smart SK, Corder GD, McFarland EW (2016) Where next on electronic waste in Australia? Waste Manag 58:348–358

Grant K, Goldizen FC, Sly PD et al (2013) Health consequences of exposure to electronic waste: a systematic review. Lancet Glob Health 1:350–361

Guo H, Gong Y, Gao S (2010a) Preparation of high strength foam glass–ceramics from waste cathode ray tube. Mater Lett 64:997–999

Guo QJ, Yue XH, Wang MH et al (2010b) Pyrolysis of scrap printed circuit board plastic particles in a fluidized bed. Powder Technol 198:422–428

Hall WJ, Williams PT (2007) Analysis of products from the pyrolysis of plastics recovered from the commercial scale recycling of waste electrical and electronic equipment. J Anal Appl Pyrolysis 79:375–386

Heeks R, Subramanian L, Jones C (2015) Understanding electronic waste Management in Developing 727 countries: strategies, determinants, and policy implications in the Indian ICT sector. Information Technol Develop 21:653–667

Herat S (2008) Recycling of cathode ray tubes (CRTs) in electronic waste. Clean (Weinh) 36:19–24

IBT (International Business Times), 2012. Electronic waste rich in silver and gold, but most unrecovered, Experts Say. 6 Jul, 12, from: http://www.ibtimes.com/electronic waste-rich-silver-and-gold-most-unrecovered-experts-say-721602

International Telecommunication Union (2017) Status of the transition to Digital Terrestrial Television Broadcasting, from: http://www.itu.int/en/ITUD/SpectrumBroadcasting/Pages/DSO/Default.aspx

International Telecommunication Union–Radiocommunication Sector (2015) ITU-R FAQ on the Digital Dividend and the Digital Switchover, from: http://www.itu.int/en/ITU-R/Documents/ITUR-FAQ-DD-DSO.pdf

Kumar A, Holuszkoa M, Espinosa DCR (2017) Electronic waste: an overview on generation, collection, legislation and recycling practices. Resour Conserv Recycling 122:32–42

Li J, Zeng X, Chen M et al (2015) Control-alt-delete rebooting solutions for the electronic waste problem. Environ Sci Technol 49:7095–7108

Lohse J, Winteler S, Wulf-Schnabel J (1998) Collection targets for waste from electrical and electronic equipment (WEEE) the directorate general (DG XI) environment. Nuclear safety and civil protection of the Commission of the European Communities

Needhidasan S, Samuel M, Chidambaram R (2014) Electronic waste – an emerging threat to the environment of urban India. J Environ Health Sci Eng 12:36

Osibanjo O, Nnorom IC (2007) The challenge of electronic waste (electronic waste) management in developing countries. Waste Manage Res 25:489–501

Siddiqi MA, Ronald H. Laessig RH et al (2003) Polybrominated diphenyl ethers (PBDEs): new pollutants-old diseases. Clin Med Res 1(4): 281–290

Step Initiative (2014) One Global Definition of Electronic waste. United Nations University, Bonn, Germany, Retrieved from http://www.step-initiative.org/files/step/documents/StEP WP OneGlobalDefinitionofElectronic waste 20140603 amended.pdf

Tsydenova O, Bengtsson M (2011) Chemical hazards associated with treatment of waste electrical and electronic equipment. Waste Manag 31(1):45–58

UNEP Blog (2014) Electronic waste, the fastest growing waste stream in the world available at. http://www.unep.org/unea/e_waste.asp

United Nations University, The Global Electronic waste Monitor 2014: Quantities, Flows and Resources, From: http://i.unu.edu/media/unu.edu/news/52624/UNU-1stGlobal-Electronic waste-Monitor-2014-small.pdf [accessed Jan 08 2016]

Veit HM, Bernardes AM (2015) Electronic waste. Springer

Yi R, Chen X, Shen L, Hu J (2007) Study on the equipment for automatic dismantling of electronic components in printed circuit boards scrap. Agricultural Equipment & Vehicle Eng 9:46–48. (in Chinese)

Zhou L, Xu Z (2012) Response to waste electrical and electronic equipments in China: legislation, recycling system, and advanced integrated process. Environ Sci Technol 46:4713–4724

Chapter 12
E-waste and Their Implications on the Environment and Human Health

Barkha Vaish, Bhavisha Sharma, Pooja Singh, and Rajeev Pratap Singh

Contents

Abstract Rapid influx of modern technology in the past few decades has led to an exponential increase in the usage of the electrical and electronic equipment on a global level. This unprecedented increase, on one hand, has revolutionized the field of communication and information technology, providing a major boost to business and economic activities; however, it has also led to the generation of one of the fastest-growing waste streams in the world, popularly referred to as E-waste. Constituents of E-waste are both hazardous and nonhazardous and valuable, comprising of toxic elements (Cd, Cr, Hg, As, Pb, Se), radioactive active substances, halogenated compounds (polychlorinated biphenyls, polybrominated biphenyls, polybrominated diphenyl ethers, chlorofluorocarbon, etc.), plastics, glass, ceramics, rubber, ferrous and non-ferrous metals (Al, Cu) and precious metals like Au, Ag, and Pt. With 20–50 million tonnes of global E-waste generation and an anticipated growth of 33%, the problem of rapidly growing E-waste is an issue faced by both developed and developing countries of the world. Additionally, unscientific and crude disposal and recycling practices for management of E-waste have severe

B. Vaish · B. Sharma · R. P. Singh (✉)
Department of Environment and Sustainable Development, Institute of Environment and Sustainable Development, Banaras Hindu University, Varanasi, India

P. Singh
Department of Science, Institute of Computer Science and Technology, SHEPA, Varanasi, India

A. Khan et al. (eds.), *E-waste Recycling and Management*, Environmental Chemistry for a Sustainable World 33,
https://doi.org/10.1007/978-3-030-14184-4_12

implications for the environment and human health resulting from release and exposure to toxic emissions and constituents. In view of the above, the present chapter attempts to provide a brief insight on the global trends of E-waste generation, critical issues and challenges associated with E-waste and its effects on environmental and human health, thereby highlighting the need for sustainable environmental management of this newer waste stream.

Keywords E-waste · Heavy metal toxicity · Environment · Human health · Sustainable waste management

12.1 Introduction

Technological revolution and advancement in the field of the electrical and electronic equipment are unprecedented in the present twenty-first century. Evidently, the developed nations of the world use more than 900 different types of electronic and electrical goods and gadgets (Huisman et al. 2012). Tremendous growth in the electronics market resulting from higher consumer demand and relatively short useful life of equipment/gadgets especially personal computers, laptops, tablets, smartphones, television sets, kitchen appliances, etc. has led to generation of a newer and bigger waste stream comprising of outdated electronic goods popularly referred to as E-waste, i.e. Electronic waste (Wong et al. 2007; Nnorom and Osibanjo 2008; Dwivedy and Mittal 2010). With the growth rate of 4% per year, the world currently generates nearly 50 million tons of E-waste annually, making it one of the fastest-growing streams of solid waste (UNEP 2005; Wang et al. 2013). According to an estimate, more than 130 million computers and television sets become obsolete in the USA on an annual basis. During the years 1997–2007, more than 500 million computers in the USA and around 610 million till 2010 in Japan were discarded (Bushehri 2010). E-waste is a complex category of solid waste containing both valuable and hazardous substances such as plastics, precious and non-precious metals (Au, Ag, Pd, Pt, Fe, Cu, Al, etc.), Pb-containing glass, Hg, Cd-containing batteries, toxic organics, flame retardants and chlorofluorocarbons (Wang et al. 2012). Improper handling and management of E-wastes may result in loss of resources and also cause environmental damage (Table 12.1).

Burgeoning amounts of E-waste generation and its hazardous constituents, trans-boundary movement and disposal issues have aroused significant environmental concerns worldwide and especially in developing countries pertaining to their growing consumption rates and imports from developed nations. Nearly 80% of the electronic goods or E-waste rejected from the developed countries on account of being old and less eco-friendly is carried across the developing nations (Hicks et al. 2005). Lack of proper legal framework and policies regarding safe disposal of imported E-waste in developing countries of Asia and Africa further intensifies the problem with serious implications to the environment and human health (Kiddee et al. 2013). Unregulated and primitive techniques of E-waste recycling such as

Table 12.1 Comparing E-waste generated and heavy metal concentration between India, China and Nigeria

	India	China	Nigeria
E-waste generated	1.7 Mt	6.0 Mt	0.22 Mt
Rate of recycling	5%	34.6%	ND
Air	Cr, 18; Mn, 59.6; Cu, 111; Zn, 191; Mo, 81.6 ng/m^3	*Mechanical workshop*: Cr, 0.554; Cu, 27.76; Cd, 0.108; Pb, 12.34 mg/g (manual)	ND
		Workshop: Cr, 0.436; Cu, 31.80; Cd, 0.398; Pb, 02.043 mg/g	
Water	*Waste leachate*: Al, 1315; Cd <1; Cu, 185; Ni, 9; Pb, 4; Zn, 17 ppm	*Well*: Cd, 5.60; Cr, 0.058; Cu, 112; Mn, 138; Ni, 3.07; Pb, 1.37	*Well*: Pb, 1.8; Cd, 0.006; Zn, 0.84; Cr, 0.25; Ni, 1.23
Soil	*DS*: Cr, 73; Cd, 2.33; Cu, 592; Mn, 449; Pb, 297; Zn, 326	*DS*: Cd, 52; Cr, 2.51; Cu, 107; Mn, 1.01; Ni, 2.52; Pb, 111; Zn, 5.40	*DS*: Pb, 502; Cd, 7.82; Zn, 66.9; Cr, 32.65; Ni, 84.24
	RS: Cr, 54; Cd, 0.47; Pb, 126; Mn, 619; Zn, 129; Cu, 429 μg/g	*BS*: Cd, 195; Cr, 3.45; Cu, 413; Mn, 1.12; Ni, 2.89; Pb, 115; Zn, 5.40	
Human health	Cr, 0.29; Mn, 1.16; Cu, 23; Zn, 141; Mo, 0.041; Ag, 2.1 μg/g	As, 0.282; Cd, 0.209; Cr, 1.16; Cu, 10.2; Mn, 1.03; Ni, 0.812; Pb, 2.98 mg/g	ND

Adapted from: Awasthi et al. (2016)
ND no data, *DS* dumping site, *RS* recycling site, *BS* burning site
(a) Balde et al. (2015), (b) Ha et al. (2009), (c) Jha et al. (2011), (d) Wu et al. (2015), (e) Wang et al. (2009), (f) Fang et al. (2013), and (g) Olafisoye et al. (2013)

dismantling, burning, roasting, melting and acid bath are hugely popular in the illegal workshops and factories in the developing countries because of their operational ease and low cost (Ren et al. 2014). These processes/methods, however, contribute immensely to the pollution of surrounding aquatic and terrestrial ecosystems and atmosphere (Fu et al. 2008). Emission or formation of highly toxic metals (Cr, Cd, Pb, Hg, Li, Be, Ba, etc.) and pollutants like polyhalogenated organics including polychlorinated biphenyls, polybrominated diphenyl ethers, polychlorinated dibenzo-p-dioxins and dibenzo-furans occurs during recycling of E-waste and causes pollution of the ambient environment (Wang et al. 2005; Sharma et al. 2018). In a study conducted by Liu et al. (2008), soil, biota and plant samples from an E-waste recycling site in South China showed contamination with significantly higher levels of polychlorinated biphenyls, polybrominated diphenyl ethers and polychlorinated dibenzo-p-dioxins and dibenzo-furans as compared to control sites. Sjödin et al. (1999) also found significantly higher levels of polybrominated diphenyl ethers in the serum of workers at E-waste recycling factories. Adverse effects of E-waste recycling are evident on human health also. Constant exposure of workers and local inhabitants to the toxic chemicals at these sites leads to their bioaccumulation in the body tissues and biomagnification through the food chain (Wong et al. 2007).

12.2 Trends in E-waste Generation

With a largely growing competitive global market of electronic and electrical products and their shorter useful life, E-waste has become a worldwide phenomenon with a growth rate of 5–10% per year (Zheng et al. 2013; Sthiannopkao and Wong 2013). Estimates provided by solving the E-waste problem (StEP) initiative predicted global production of E-waste in the world to rise from 49 million tonnes during 2012 to around 65.4 million tonnes till 2017 (UNU 2013). Higher generation rates and potential hazardous impacts on the environment and human health caused by associated toxic chemicals during recycling and disposal have made sustainable E-waste management a major environmental concern and challenge (Leung et al. 2007; Wu et al. 2008; Luo et al. 2009). Majority of the E-waste from developed world, i.e. the USA, Britain and Europe, is transferred to developing nations like China, India and Nigeria (Chi et al. 2011). E-waste management is an issue dealt differently in terms of policies and methodologies adopted in both developed and developing countries of the world. Developed countries have expensive, fairly regulated and well-devised collection systems, clean recovery technologies such as disassembly stations, and plasma furnaces installed to prevent toxic emissions from E-waste recycling; however still, the majority of the European and North American E-waste remains unrecycled (Barba-Gutiérrez et al. 2008). On the other hand, most of the developing countries (like China, India, Pakistan, Indonesia, the Philippines, Nigeria, etc.) in absence of proper laws, policies and regulatory guidelines resort to primitive, cheap and crude recycling practices such as smouldering, acid baths, crushing, open incineration, etc. at the informal recycling facilities to manage E-waste causing severe damage to the environment and human health (Widmer et al. 2005; Zheng et al. 2013; Yoshida et al. 2016). During the year 2010, developed countries like Japan and the European Union generated 4 and 8.9 million tonnes of domestic E-waste, respectively (Zoeteman et al. 2010), and around 400 million electronic items become obsolete in the USA alone per year. According to Widmer et al. (2005), in the developed countries, E-waste may constitute 8% by volume of municipal solid waste. Receiving more than 70% of the E-waste in the world and also being its second largest producer, China is likely to overtake the USA's E-waste generation by the year 2020 (Hicks et al. 2005; UNEP 2007). It is predicted that by 2030, developing countries will discard approximately 600 million personal computers per year which is twice the number that will be discarded by the developed nations, i.e. 300 million (Yu et al. 2010). In economically advanced Southeast Asian countries like Malaysia and Thailand, E-waste generation was estimated to be 6–10 kg per capita during 2012, whereas it ranged from 2 to 3 kg per capita for middle-income countries like the Philippines, Vietnam and Indonesia (Yoshida et al. 2016).

12.3 Environmental Implication of E-waste

In developing nations, widespread unscientific methods of E-waste recycling have caused severe contamination of soil, water, air and adverse health effects as these products have a high concentration of harmful materials like heavy metals and persistent organic pollutants. E-wastes comprise different materials; most of them are toxicant and cause serious environmental problems (Pant et al. 2012; Chatterjee 2008). Two major kinds of substances are released from E-waste, viz. hazardous (heavy metals, polybrominated diphenyl ethers, polycyclic aromatic hydrocarbons and polychlorinated dibenzo-p-dioxins and dibenzofurans) and nonhazardous (metals like Cu, Zn and Se along with precious metals like Au, Ag, Pt, etc.) (Awasthi et al. 2016; Wei et al. 2014; Zeng 2014; Zhang et al. 2013). Both types of E-waste cause potential harmful impact on the environment if present beyond their permissible limit (Pant 2010). During processing of these E-wastes, a significant amount of toxic compounds are released which causes a detrimental effect on human well-being and the surrounding environment (Robinson 2009; Shen et al. 2009).

Additionally, around 80% of E-waste is illegally transported from advanced to developing nations like India, China, Ghana, Nigeria, and Pakistan because of governmental negligence, lack of stringent policies and lower cost of labour (Pradhan and Kumar 2014; Sthiannopkao and Wong 2013). Several scientists have investigated that workshop sites for processing/transforming of E-waste release toxic substances that contaminate natural surrounding environment (Kwatra et al. 2014; Wu et al. 2014; Stevels et al. 2013). In continuation, many scientists have documented heavy metal contamination in the surrounding environment (water, air and soil) near recycling workshops mainly in developing nations (Leung et al. 2006, 2007, 2008; Wong et al. 2007).

Leaching is the process through which harmful contaminants enter aquatic systems where processed and unprocessed waste may be deposited. Likewise, settling/dissolution of airborne contaminants or hydrometallurgical processes followed by acid disposal penetrate into water or soil thus contaminating soil and aquatic systems. However, e-waste contaminants enter human body via skin absorption, inhalation and ingestion by dust-laden air (Mielke and Reagan 1998). A parallel study by Ha et al. (2009) indicated that dust-laden air near E-waste processing workshops in Bangalore contains contaminants like Sn, Pb, Sb, Cd, In and Bi at a high level of 91, 89, 13, 1.5, 1.3 and 1.0 ng m^{-3}, respectively. Most of the E-waste in developing nations are dealt with little concern often by acid baths and open burning for recovery of a few valuable materials. Result of these processes are harmful contaminants like polybrominated diphenyl ethers, polychlorinated biphenyls, heavy metals, dioxins, furans are observed in surrounding environments, workers and nearby residents. Hence, the environmental impact of the pollutants (polychlorinated biphenyls, polybrominated diphenyl ethers, heavy metals) must be discussed extensively regarding the health and safety of workers or the environment.

12.3.1 Heavy Metal Toxicity

Manufacturing of electronic devices widely uses heavy metals like Cd and Pb in circuit boards and computer batteries, Cu in electrical wiring, etc. (Achillas et al. 2013; Stevels et al. 2013; Zeng et al. 2014). A study done by Morf et al. (2007) found that the average fraction of plastic in E-waste has Pb, Ni, Sn, Zn and Sb at a concentration > 1000 mg kg^{-1} and > 100 mg kg^{-1} for Cd. Primitive processes and techniques are extremely popular in developing nations like India and China. Therefore, it became a new cause of environmental pollution these decades (Chi et al. 2011; Song and Li 2014). The unregulated processing by using primitive techniques like a chemical process, acid baths and burning to procure valuable metals cause severe heavy metal pollution in the terrestrial and aquatic ecosystem (Deng et al. 2007; Wei and Liu 2012).

Additionally, incineration done prior to landfilling increases the mobility of heavy metals, especially Pb (Gullett et al. 2007). In a study by Luo et al. (2011), which collected samples of soils and vegetables from prior incineration sites for heavy metal analysis, the results demonstrated high doses of Zn, Pb, Cu and Cd with values of 3690, 4500, 11,140 and 17.1 mg kg^{-1}, respectively. Also, soils of paddy fields and nearby gardens had a comparatively high level of Cu and Cd. Similarly, the high level of Cd and Pb was observed in inedible parts of vegetables, which goes beyond the maximum permissible limit in China. High doses of Cd, Pb and polybrominated biphenyls were found higher in rice and other crops cultivated near E-waste recycling units. This is because heavy metals are discharged during recovery of precious metals that enter the soils where crops and vegetables are grown by polluting irrigation water or through foliar uptake of heavy metals by air (Bi et al. 2009). Oral intake of contaminated food is an important pathway for translocation of heavy metals from the environment to the human population.

Developing nations like China and India are a hub for recovery of precious metals from E-waste by informal reprocessing of printed circuit boards, batteries and cables. The complete step of the process is accomplished by people involved irrespective of gender and age who work without using proper protection in a harmful environment. In a study by Ha et al. (2009) observed high doses of heavy metal in soil at a recycling slum in Bangalore that contained up to 4.6 mg kg^{-1} In, 180 mg kg^{-1} Sb, 2850 mg kg^{-1} Pb, 39 mg kg^{-1} Cd, 49 mg kg^{-1} Hg, 957 mg kg^{-1} Sn and 2.7 mg kg^{-1} Bi. The concentrations recorded were 100 times more than control site. A similar study by Pradhan and Kumar (2014) analysed heavy metals in soil, water and plant samples collected from recycling sites in Mandoli Industrial Area, Delhi. Results revealed that a high level of heavy metals concentration was found in soil samples like Cu (115.50 mg kg^{-1}), As (17.08 mg kg^{-1}), Pb (2645.31 mg kg^{-1}), Cd (1.29 mg kg^{-1}), Se (12.67 mg kg^{-1}), Zn (776.84 mg kg^{-1}), native plant samples (*Cynodon dactylon*) and water samples.

12.3.2 Hazardous Chemical Toxicity

E-waste is different from other forms of wastes which comprise composite mixtures of probable environmental contaminants. E-waste also contains some uncommon potential contaminants even in other polluted sites. In the manufacturing of electrical items, some heavy metals contaminants are used, while during combustion of E-waste at low temperature, other contaminants such as polycyclic aromatic hydrocarbons are produced. Important metals (Cu, platinum group) along with potential environmental contaminants, especially Ni, Sb, Hg, Pb, Cd, polybrominated diphenyl ethers and polychlorinated biphenyls, are generally present in the E-waste. Dioxins, furans, polycyclic aromatic hydrocarbons, polybrominated diphenyl ethers, polychlorinated biphenyls and polychlorinated dibenzo-p-dioxins and dibenzofurans and hydrogen chloride could be generated due to the burning of E-waste that are highly contaminated pollutants (Darnerud et al. 2001; Martin et al. 2003).

Open burning and labour-intensive treatment are widely used basic methods in the recycling of E-waste. In comparison with domestic waste, ignition of insulated wire produced 100X more dioxins (Gullett et al. 2007). Flame retardants such as polybrominated diphenyl ethers are synthesized into plastics and other components. Polybrominated diphenyl ethers and plastics can leach out from the surface of E-waste into the atmosphere due to the absence of chemical bonds between them (Deng et al. 2007). Lipophilic character of polybrominated diphenyl ethers causes bioaccumulation in the organisms and biomagnification in the food chains (Deng et al. 2007). Endocrine disrupting properties are also found in polybrominated diphenyl ethers (Tseng et al. 2008). During the recycling procedures, ignition products of polyvinyl chloride in electrical wires can emit or form a number of highly lethal pollutants like polychlorinated biphenyls in capacitors/transformers, polybrominated diphenyl ethers which are used as brominated flame retardants in the circuit board and polychlorinated dibenzo-p-dioxins and dibenzofurans (Wang et al. 2005). Soil samples were gathered by Leung et al. (2007) from the site where it was usual to liquify circuit boards, incinerate things as cable covering to recover copper wire and employ open-pit acid leaching for extraction of valuable metals. Their research works concentrated on polybrominated diphenyl ethers and polychlorinated dibenzo-p-dioxins along with dibenzofurans, that is, polybrominated diphenyl ethers and polychlorinated dibenzo-p-dioxins and dibenzofurans. Carcinogens in duck ponds and paddies were 263–604 ng g^{-1}, dry weight and 34.7–70.9 ng g^{-1} which exceeded from control sites.

A similar study by Luo et al. (2007b) accounted that carps collected from the Nanyang River, near Guiyu, bioaccumulated polybrominated diphenyl ethers to high level, i.e. 766 ng g^{-1} (fresh weight). Expectedly, Luo et al. (2007a) in the further study accounted for high polybrominated diphenyl ether levels in the sediments, i.e. up to 16,000 ng g^{-1} of the Nanyang River. A similar study was done by Wu et al. (2008) and found that water snake near (the top predator) an E-waste recycling yard had around 1091 ng g^{-1} polybrominated diphenyl ethers and

16,512 ng g^{-1} polychlorinated biphenyls on a wet weight basis. Other than polybrominated diphenyl ethers, brominated flame retardants like decabromodiphenyl ethane, tetra-bromobisphenol A bis (2, 3-dibromopropyl) ether and 1, 2-bis (2, 4, 6-tribromophenoxy) ethane are commonly found in the different ecosystem of Pearl River Delta (Shi et al. 2009). All these have caused a high level of toxic pollutants in the ambient atmosphere that further degrade the ecosystem and human well-being (Wong et al. 2007; Yu et al. 2006; Deng et al. 2006).

12.4 Effect on Human Health

Hazardous waste may adversely affect the health of the local inhabitants and workers that may involve any organ failure depending on the contact of any specific type of chemical(s), exposure time, exposed individual's characteristics like age and sex, body weight, immunological status, etc. Exposure routes may differ depending on the kind of substance involved and their recycling process. The general route followed by the harmful components emerging from E-waste is either via inhalation, ingestion or dermal contact. In addition to this, people can come in contact with associated pollutants through contaminated air, soil, water or food. Also, additional exposure risks are faced by pregnant women, foetuses, children, elderly population, disabled persons, workers and local residents (Grant et al. 2013). Among them, children are at a higher risk due to other routes of exposure like placental exposure or breastfeeding, their altering physiology like high consumption of water and food and low toxin elimination rate and high risk-taking behaviours like hand to mouth activity in recent years (Pronczuk de Garbino 2004). Besides, the children of E-waste manufacturing and recycling workers get contaminated by their parents' skin or clothes and direct exposure if the recycling process is happening in nearby places or homes.

Heavy metals penetrate the human body via oral intake, dermal contact and inhalation. Water and food intake are the major sources of oral exposure to heavy metals (Zheng et al. 2013; Xu et al. 2006). Food crops bioaccumulate heavy metals via wastewater irrigation (Singh et al. 2010), atmospheric deposition (Bi et al. 2009) or contaminated soil (Zhuang et al. 2009). Contaminated soil and feeds result in an elevated level of heavy metals in meat products (Cang et al. 2004; Gonzalez-Weller et al. 2006). A number of studies revealed that Pb, Cd, Zn and Cu are potential human carcinogens that are related to several disorders like the nervous system, blood, urine, cardiovascular and bone diseases (Jarup 2003; Brewer 2010; Muyssen et al. 2006). A similar study by Thomas et al. (2009) speculated that Cd and other heavy metals cause early kidney damage. Several studies have also found that Cu leads to liver damage, Pb results in behaviour and learning disabilities and Cd increases the risk of kidney damage (Bhutta et al. 2011; Chan et al. 2013; Yan et al. 2013). These studies emphasize the significance of assessment of heavy metal risk and exposure to the local inhabitants and labours in E-waste recycling areas.

Table 12.2 Health effects and exposure route of different pollutant emitted from E-waste

S. no.	Pollutant	Exposure route	Effects
1.	Heavy metals	Air, dust, water, soil, food	Human carcinogen; affect neurodevelopmental activity, cognition, learning and behaviour; and affect neuromotor skills
2.	PBDEs	Air, dust, food	Thyroid hormone disruption, hyperactivity, cognitive deficits and impaired memory
3.	PCBs	Air, dust, seafood	Affect neuropsychological functions in children, including general cognition, visual-spatial function, memory, attention, executive functions and motor function
4.	PCDD/ PCDFs	Air, dust, soil, food	Reproductive and neurobehavioural development, immune development, carcinogenicity
5.	PAHs	Air, dust, soil, food	Carcinogen and mutagen affect child neurodevelopment and lead to IQ deficits

A comparable study by Ha et al. (2009) in recycling sites such as Guiyu in China and Bangalore in India speculated that heavy metals and their associated pollutants cause severe damage to the environment and human health. This is due to a high penetration rate of heavy metals in soil and then to plants where it bio-accumulates and further transported to trophic level. They found an elevated level of heavy metals like Pb, Bi, Cu, Zn, Bi, In and Sn in the soil near recycling workshops. Hair samples of workers have increased the level of Cd, Ag, Cu, Sb and Bi. As compared to control, ten times higher level of polybrominated diphenyl ethers was found in the serum of workers at recycling workshops (Sjödin et al. 1999). Also, the children and neonates have elevated levels of Cr, Ni, Cd and polybrominated diphenyl ethers than controls (Guo et al. 2010; Wu et al. 2010). A parallel study by Asante et al. (2012) at Agbogbloshie, Accra and Ghana found relatively high traces of heavy metals in the urine of workers. These studies indicated that E-waste disposal has a serious negative impact on human health and the environment (Table 12.2).

12.5 Conclusion

Already a global concern regarding E-waste management has resulted in many policies and legislation. The European Union has implemented two such legislations to address the concern over the management of E-waste. The first directive is on E-waste management, i.e. waste electrical and electronic equipment, which directs the manufacturer about their responsibility for management of E-waste; this strategy is called greener electronics (Chen et al. 2011; Pant et al. 2012). Similarly, 'Restriction on the use of Hazardous Substances (RoHS) Directive' limits the application of heavy metals like Cd, Pb, Hg, Cr (VI), polybrominated diphenyl ethers and polybrominated biphenyls in the manufacture of new devices (Chen et al. 2011).

Likewise, India drafted 'Hazardous Materials Laws and Rules' in 2007 for addressing the same issue (LaDou and Lovegrove 2008).

Non-governmental organizations have played a crucial role by placing pressure on manufacturers to eliminate or limit the use of environmental contaminants during manufacturing of their products. One such concept is extended producer responsibility (EPR) that provides incentives for redesigning and removing toxic contaminants from their products (Betts 2008). Many manufacturers of electronic goods have started exploring innovative measures to improve recycling and safe disposal of E-waste. Developed countries have the self-interest to mitigate the harmful environmental effects of E-waste as it will negatively affect the food quality and quantity and other goods that are manufactured and imported from developing countries. However, there is limited information on environmental effects, risks on human health and remediation techniques for most of the E-waste contaminants; therefore, it is anticipated that more safe and scientific recycling options of E-waste will be adopted to avoid damage of the local environment and human health.

Acknowledgement The authors are thankful to the Dean & Head, Department of Environment and Sustainable Development and Director, Institute of Environment and Sustainable Development, Banaras Hindu University, for providing necessary facilities. RPS is thankful to the Department of Science and Technology for providing financial support (DST-SERB P07-678). BS is thankful to the University Grants Commission for awarding Junior and Senior Research Fellowship. BV is also thankful to the Council of Scientific and Industrial Research for awarding Senior Research Fellowship.

Conflict of Interest On behalf of all authors, the corresponding author states that there is no conflict of interest.

References

Achillas C, Aidonis D, Vlachokostas C, Karagiannidis A, Moussiopoulos N, Loulos V (2013) Depth of manual dismantling analysis: a cost-benefit approach. Waste Manag 33:948–956

Asante KA, Agusa T, Biney CA, Agyekum WA, Bello M, Otsuka M, Itai T, Takahashi S, Tanabe S (2012) Multi-trace element levels and arsenic speciation in urine of e-waste recycling workers from Agbogbloshie, Accra in Ghana. Sci Total Environ 424:63–73

Awasthi AK, Zeng X, Li J (2016) Environmental pollution of electronic waste recycling in India: a critical review. Environ Pollut 211:259–270

Balde CP, Wang F, Kuehr R, Huisman J (2015) The global E-waste monitor 2014. Quantities flows and resources. United Nations University, IAS e SCYCLE, Bonn, Germany, pp 1–41. Institute for the Advanced Study of Sustainability. http://i.unu.edu/media/ias.unu.edu-en/news/7916/Global-E-waste-Monitor-2014-small.pdf

Barba-Gutiérrez Y, Adenso-Diaz B, Hopp M (2008) An analysis of some environmental consequences of European electrical and electronic waste regulation. Resour Conserv Recycl 52(3):481–495

Betts K (2008) Reducing the global impact of e-waste. Environ Sci Technol 42:1393–1393

Bhutta MKS, Omar A, Yang X (2011) Electronic waste: a growing concern in today's environment. Econ Res Int 2011:1–8

Bi XY, Feng XB, Yang YG, Li XD, Shin GPY, Li FL, Qiu GL, Li GH, Liu TZ, Fu ZY (2009) Allocation and source attribution of lead and cadmium in maize (*Zea mays* L.) impacted by smelting emissions. Environ Pollut 157:834–839

Brewer GJ (2010) Copper toxicity in the general population. Clin Neurophysiol 121:459–460

Bushehri FI (2010) UNEP's role in promoting environmentally sound management of e-waste. In: 5th ITU symposium on "ICTs, the environment and climate change". Cairo, Egypt

Cang L, Wang YJ, Zhou DM, Dong YH (2004) Heavy metals pollution in poultry and livestock feeds and manures under intensive farming in Jiangsu Province. China J Environ Sci 16:371–374

Chan JK, Man YB, Wu SC, Wong MH (2013) Dietary intake of PBDEs of residents at two major electronic waste recycling sites in China. Sci Total Environ 463–464:1138–1146

Chatterjee P (2008) Health costs of recycling. Br Med J 337:376–377

Chen A, Dietrich KN, Huo X, Ho S-M (2011) Developmental neurotoxicants in e-waste: an emerging health concern. Environ Health Perspect 119:431–438

Chi X, Streicher-Porte M, Wang MY, Reuter MA (2011) Informal electronic waste recycling: a sector review with special focus on China. Waste Manag 31:731–742

Darnerud PO, Eriksen GS, Jóhannesson T, Larsen PB, Viluksela M (2001) Polybrominated diphenyl ethers: occurrence, dietary exposure, and toxicology. Environ Health Perspect 109:49–68

Deng WJ, Louie PKK, Liu WK, Bi XH, Fu JM, Wong MH (2006) Atmospheric levels and cytotoxicity of PAHs and heavy metals in TSP and PM2.5 at an electronic waste recycling site in southeast China. Atmos Environ 40:6945–6955

Deng WJ, Zheng JS, Bi XH, Fu JM, Wong MH (2007) Distribution of PBDEs in air particles from an electronic waste recycling site compared with Guangzhou and Hong Kong, South China. Environ Int 33:1063–1069

Dwivedy M, Mittal RK (2010) Estimation of future outflows of e-waste in India. Waste Manag 30(3):483–491

Fang WX, Yang YC, Xu ZM (2013) PM10 and PM2.5 and health risk assessment for heavy metals in a typical factory for cathode ray tube television recycling. Environ Sci Technol 47(21):12469–12476. https://doi.org/10.1021/es4026613

Fu J, Zhou Q, Liu J, Liu W, Wang T, Zhang Q, Jiang G (2008) High levels of heavy metals in rice (Oryza sativa L.) from a typical E-waste recycling area in southeast China and its potential risk to human health. Chemosphere 71(7):1269–1275

Gonzalez-Weller D, Karlsson L, Caballero A, Hernandez F, Gutierrez A, Gonzalez-Iglesias T, Marino M, Hardisson A (2006) Lead and cadmium in meat and meat products consumed by the population in Tenerife Island, Spain. Food Addit Contam 23:757–763

Grant K, Goldizen FC, Sly PD, Brune MN, Neira M, van den Berg M, Norman RE (2013) Health consequences of exposure to e-waste: a systematic review. Lancet Glob Health 1(6):e350–e361

Gullett BK, Linak WP, Touati A, Wasson SJ, Gatica S, King CJ (2007) Characterization of air emissions and residual ash from open burning of electronic wastes during simulated rudimentary recycling operations. J Mater Cycles Waste Manag 9:69–79

Guo Y, Huo X, Li Y, Wu K, Liu J, Huang J, Zheng G, Xiao Q, Yang H, Wang Y, Chen A, Xu X (2010) Monitoring of lead, cadmium, chromium and nickel in placenta from an e-waste recycling town in China. Sci Total Environ 408:3113–3117

Ha NN, Agusa T, Ramu K, Tu NPC, Murata S, Bulbule KA, Parthasaraty P, Takahashi S, Subramanian A, Tanabe S (2009) Contamination by trace elements at e-waste recycling sites in Bangalore, India. Chemosphere 76(1):9–15

Hicks C, Dietmar R, Eugster M (2005) The recycling and disposal of electrical and electronic waste in China—legislative and market responses. Environ Impact Assess Rev 25(5):459–471

Huisman J et al (2012) The Dutch WEEE flows. United Nations University, ISPSCYCLE, Bonn

Jarup L (2003) Hazards of heavy metal contamination. Br Med Bull 68:167–182

Jha MK, Kumar A, Kumar V, Lee JC (2011) Prospective scenario of e-waste recycling in India. In: Recycling of electronic waste II: proceedings of the second symposium, pp 73–80. https://doi.org/10.1002/9781118086391.ch10

Kiddee P, Naidu R, Wong MH (2013) Electronic waste management approaches: an overview. Waste Manag 33(5):1237–1250
Kwatra S, Pandey S, Sharma S (2014) Understanding public knowledge and awareness on e-waste in an urban setting in India. A case study for Delhi. Manag Environ Qual 25(6):752–765
Ladou J, Lovegrove S (2008) Export of electronics equipment waste. Int J Occup Environ Health 14:1–10
Leung A, Cai ZW, Wong MH (2006) Environmental contamination from electronic waste recycling at Guiyu, southeast China. J Mater Cycles Waste Manag 8(1):21–33
Leung AOW, Luksemburg WJ, Wong AS, Wong MH (2007) Spatial distribution of polybrominated diphenyl ethers and polychlorinated dibenzo-p-dioxins and dibenzofurans in soil and combusted residue at Guiyu, an electronic waste recycling site in southeast China. Environ Sci Technol 41:2730–2737
Leung AOW, Duzgoren-Aydin NS, Cheung KC, Wong MH (2008) Heavy metals concentrations of surface dust from e-waste recycling and its human health implications in southeast China. Environ Sci Technol 42(7):2674–2680
Liu H, Zhou Q, Wang Y, Zhang Q, Cai Z, Jiang G (2008) E-waste recycling induced polybrominated diphenyl ethers, polychlorinated biphenyls, polychlorinated dibenzo-p-dioxins and dibenzofurans pollution in the ambient environment. Environ Int 34(1):67–72
Luo Q, Cai ZW, Wong MH (2007a) Polybrominated diphenyl ethers in fish and sediment from river polluted by electronic waste. Sci Total Environ 383:115–127
Luo Q, Wong MH, Cai ZW (2007b) Determination of polybrominated diphenyl ethers in freshwater fishes from a river polluted by e-wastes. Talanta 72:1644–1649
Luo XJ, Liu J, Luo Y, Zhang XL, Wu JP, Lin Z, Chen SJ, Mai BX, Yang ZY (2009) Polybrominated diphenyl ethers (PBDEs) in free-range domestic fowl from an e-waste recycling site in South China: levels, profile and human dietary exposure. Environ Int 35(2):253–258
Luo C, Liu C, Wang Y, Liu X, Li F, Zhang G, Li X (2011) Heavy metal contamination in soils and vegetables near an e-waste processing site, south China. J Hazard Mater 186(1):481–490
Martin M, Richardson BJ, Lam PKS (2003) Harmonisation of polychlorinated biphenyl (PCB) analyses for ecotoxicological interpretations of Southeast Asian environmental media: what's the problem? Mar Pollut Bull 46:159–170
Mielke HW, Reagan PL (1998) Soil is an important pathway of human lead exposure. Environ Health Perspect 106:217–229
Morf LS, Tremp J, Gloor R, Schuppisser F, Stengele M, Taverna R (2007) Metals, non-metals and PCB in electrical and electronic waste—actual levels in Switzerland. Waste Manag 27:1306–1316
Muyssen BTA, De Schamphelaere KAC, Janssen CR (2006) Mechanisms of chronic waterborne Zn toxicity in *Daphnia magna*. Aquat Toxicol 77:393–401
Nnorom IC, Osibanjo O (2008) Electronic waste (e-waste): material flows and management practices in Nigeria. Waste Manag 28(8):1472–1479
Olafisoye OB, Tejumade A, Otolorin AO (2013) Heavy metals contamination of water, soil and plants around an electronic waste dumpsite. Pol J Environ Stud 22(5):1431–1439
Pant D (2010) Electronic waste management. Lambart Academic Publishing, Saarbrücken, pp 3–16. ISBN 978-3-8433-8336-3
Pant D, Joshi D, Upreti MK, Kotnala RK (2012) Chemical and biological extraction of metals present in E waste: a hybrid technology. Waste Manag 32(5):979–990
Pradhan JK, Kumar S (2014) Informal e-waste recycling: environmental risk assessment of heavy metal contamination in Mandoli industrial area, Delhi, India. Environ Sci Pollut Res 21(13):7913–7928
Pronczuk de Garbino J (2004) In: Pronczuk de Garbino J (ed) Children's health and the environment: a global perspective. A resource manual for the health sector. World Health Organization, New York

Ren Z, Xiao X, Chen D, Bi X, Huang B, Liu M, Hu J, Peng P, Sheng G, Fu J (2014) Halogenated organic pollutants in particulate matters emitted during recycling of waste printed circuit boards in a typical e-waste workshop of Southern China. Chemosphere 94:143–150

Robinson BH (2009) E-waste: an assessment of global production and environmental impacts. Sci Total Environ 408(2):183–191

Sharma B, Vaish B, Srivastava V, Singh S, Singh P, Singh RP (2018) An insight to atmospheric pollution-improper waste management and climate change nexus. In: Modern age environmental problems and their remediation. Springer, Cham, pp 23–47

Shen C, Tang X, Cheema SA, Zhang C, Khan MI, Liang F et al (2009) Enhanced phytoremediation potential of polychlorinated biphenyl contaminated soil from e-waste recycling area in the presence of randomly methylated-beta-cyclodextrins. J Hazard Mater 172:1671–1676

Shi T, Chen SJ, Luo XJ, Zhang XL, Tang CM, Luo Y et al (2009) Occurrence of brominated flame retardants other than polybrominated diphenyl ethers in environmental and biota samples from southern China. Chemosphere 74:910–916

Singh A, Sharma RK, Agrawal M, Marshall FM (2010) Health risk assessment of heavy metals via dietary intake of foodstuffs from the wastewater irrigated site of a dry tropical area of India. Food Chem Toxicol 48:611–619

Sjödin A, Hagmar L, Klasson-Wehler E, Kronholm-Diab K, Jakobsson E, Bergman A (1999) Flame retardant exposure: polybrominated diphenyl ethers in blood from Swedish workers. Environ Health Perspect 107(8):643–648

Song Q, Li J (2014) Environmental effects of heavy metals derived from the e-waste recycling activities in China: a systematic review. Waste Manag 34(12):2587–2594

Stevels A, Huisman J, Wang F, Li J, Li B, Duan H (2013) Take back and treatment of discarded electronics: a scientific update. Front Environ Sci Eng 7:475–482

Sthiannopkao S, Wong MH (2013) Handling e-waste in developed and developing countries: initiatives, practices, and consequences. Sci Total Environ 463:1147–1153

Thomas LDK, Hodgson S, Nieuwenhuijsen M, Jarup L (2009) Early kidney damage in a population exposed to cadmium and other heavy metals. Environ Health Perspect 117:181–184

Tseng LH, Li MH, Tsai SS, Lee CW, Pan MH, Yao WJ et al (2008) Developmental exposure to decabromodiphenyl ether (PBDE 209): effects on thyroid hormone and hepatic enzyme activity in male mouse offspring. Chemosphere 70:640–647

UNEP (2007) E-waste volume II: e-waste management manual. Division of Technology, Industry and Economics, International Environmental Technology Centre, Osaka. United Nations Environment Programme (UNEP)

UNEP DEWA/GRID-Europe (2005) Chapter 5: Early warning on emerging environmental threats. In: E-waste, the hidden side of IT equipment's manufacturing and use. www.grid.unep.ch

United Nations University (UNU) (2013) StEP launches interactive world E-waste map. _http://unu.edu/media-relations/releases/step-launches-interactive-world-e-waste-map.html#info_

Wang D, Cai Z, Jiang G, Leung A, Wong MH, Wong WK (2005) Determination of polybrominated diphenyl ethers in soil and sediment from an electronic waste recycling facility. Chemosphere 60(6):810–816

Wang T, Fu JJ, Wang YW, Liao CY, Tao YQ, Jiang GB (2009) Use of scalp hair as indicator of human exposure to heavy metals in an electronic waste recycling area. Environ Pollut 157(8–9):2445–2451. https://doi.org/10.1016/j.envpol.2009.03.010

Wang F, Huisman J, Meskers CE, Schluep M, Stevels A, Hagelüken C (2012) The best-of-2-worlds philosophy: developing local dismantling and global infrastructure network for sustainable e-waste treatment in emerging economies. Waste Manag 32(11):2134–2146

Wang F, Huisman J, Stevels A, Baldé CP (2013) Enhancing e-waste estimates: improving data quality by multivariate input–output analysis. Waste Manag 33(11):2397–2407

Wei L, Liu Y (2012) Present status of e-waste disposal and recycling in China. Procedia Environ Sci 16:506–514

Wei YL, Bao LJ, Wu CC, He ZC, Zeng EY (2014) Association of soil polycyclic aromatic hydrocarbon levels and anthropogenic impacts in a rapidly urbanizing region: spatial distribution, soil-air exchange and ecological risk. Sci Total Environ 473–474:676–684
Widmer R, Oswald-Krapf H, Sinha-Khetriwal D, Schnellmann M, Böni H (2005) Global perspectives on e-waste. Environ Impact Assess Rev 25(5):436–458
Wong MH, Wu SC, Deng WJ, Yu XZ, Luo Q, Leung AOW, Wong CSC, Luksemburg WJ, Wong AS (2007) Export of toxic chemicals – a review of the case of uncontrolled electronic-waste recycling. Environ Pollut 149(2):131–140
Wu JP, Luo XJ, Zhang Y, Luo Y, Chen SJ, Mai BX et al (2008) Bioaccumulation of polybrominated diphenyl ethers (PBDEs) and polychlorinated biphenyls (PCBs) in wild aquatic species from an electronic waste (e-waste) recycling site in South China. Environ Int 34:1109–1113
Wu K, Xu X, Liu J, Guo Y, Li Y, Huo X (2010) Polybrominated diphenyl ethers in umbilical cord blood and relevant factors in neonates from Guiyu, China. Environ Sci Technol 44:813–819
Wu C, Luo Y, Deng S, Teng Y, Song J (2014) Spatial characteristics of cadmium in topsoils in a typical e-waste recycling area in southeast China and its potential threat to shallow groundwater. Sci Total Environ 472:556–561. https://doi.org/10.1016/j.scitotenv.2013.11.084
Wu Q, Leung JY, Geng X, Chen S, Huang X, Li H (2015) Heavy metal contamination of soil and water in the vicinity of an abandoned e-waste recycling site: implications for dissemination of heavy metals. Sci Total Environ 506-507:217–225
Xu P, Huang SB, Wang ZJ, Lagos G (2006) Daily intakes of copper, zinc and arsenic in drinking water by population of Shanghai. China Sci Total Environ 362:50–55
Yan CH, Xu J, Shen XM (2013) Childhood lead poisoning in China: challenges and opportunities. Environ Health Perspect 121:A294
Yoshida A, Terazono A, Ballesteros FC Jr, Nguyen DQ, Sukandar S, Kojima M, Sakata S (2016) E-waste recycling processes in Indonesia, the Philippines, and Vietnam: a case study of cathode ray tube TVs and monitors. Resour Conserv Recycl 106:48–58
Yu XZ, Gao Y, Wu SC, Zhang HB, Cheung KC, Wong MH (2006) Distribution of poly-cyclic aromatic hydrocarbons in soils at Guiyu area of China, affected by recycling of electronic waste using primitive technologies. Chemosphere 65(9):1500–1509
Yu J, Williams E, Ju M, Yang Y (2010) Forecasting global generation of obsolete personal computers. Environ Sci Technol 44:3232–3237
Zeng EY (2014) Environmental challenges in China. Environ Toxicol Chem 33(8):1690–1691
Zeng X, Li J, Singh N (2014) Recycling of spent lithium-ion battery: a critical review. Crit Rev Environ Sci Technol 44:1129–1165
Zhang K, Wei YL, Zeng EY (2013) A review of environmental and human exposure to persistent organic pollutants in the Pearl River Delta. South China Sci Total Environ 463–464:1093–1110
Zheng J, Chen KH, Yan X, Chen SJ, Hu GC, Peng XW, Yuan JG, Mai B-X, Yang ZY (2013) Heavy metals in food, house dust, and water from an e-waste recycling area in South China and the potential risk to human health. Ecotoxicol Environ Saf 96:205–212
Zhuang P, McBride MB, Xia HP, Li NY, Lia ZA (2009) Health risk from heavy metals via consumption of food crops in the vicinity of Dabaoshan mine. South China Sci Total Environ 407:1551–1561
Zoeteman BC, Krikke HR, Venselaar J (2010) Handling WEEE waste flows: on the effectiveness of producer responsibility in a globalizing world. Int J Adv Manuf Technol 47(5–8):415–436

Index

C

D

E

F

A. Khan et al. (eds.), *E-waste Recycling and Management*, Environmental Chemistry for a Sustainable World 33, https://doi.org/10.1007/978-3-030-14184-4

MIX
Papier aus verantwortungsvollen Quellen
Paper from responsible sources
FSC® C105338
FSC
www.fsc.org

If you have any concerns about our products,
you can contact us on
ProductSafety@springernature.com

In case Publisher is established outside the EU,
the EU authorized representative is:
Springer Nature Customer Service Center GmbH
Europaplatz 3, 69115 Heidelberg, Germany

Printed by Libri Plureos GmbH
in Hamburg, Germany